Nanocellulose

Synthesis, Properties and Applications

Related Titles

Biopolymers Based Advanced Materials
ISBN: 978-0-6482205-4-1 (e-book)
ISBN: 978-0-6482205-5-8 (hardcover)

Functional Polymer Blends and Nanocomposites
ISBN: 978-0-6482205-6-5 (e-book)
ISBN: 978-0-6482205-7-2 (hardcover)

Functional Nanomaterials and Nanotechnologies: Applications for Energy & Environment
ISBN: 978-0-6482205-2-7 (e-book)
ISBN: 978-0-6482205-3-4 (softcover)

Advances in Polymer Technology: Material Development, Properties and Performance Evaluation
ISBN: 978-1-925823-00-4 (e-book)
ISBN: 978-1-925823-01-1 (hardcover)

Polymer Nanomaterials for Specialty Applications
ISBN: 978-1-925823-03-5 (e-book)
ISBN: 978-1-925823-04-2 (hardcover)

Advanced Materials
ISBN: 978-1-925823-05-9 (e-book)
ISBN: 978-1-925823-06-6 (hardcover)

Biofuels
ISBN: 978-1-925823-12-7 (e-book)
ISBN: 978-1-925823-13-4 (hardcover)

Liquid Crystalline Polymers
ISBN: 978-1-925823-16-5 (e-book)
ISBN: 978-1-925823-17-2 (hardcover)

Polymer Nanocomposites: Emerging Applications
ISBN: 978-1-925823-14-1 (e-book)
ISBN: 978-1-925823-15-8 (hardcover)

Nanocellulose

Synthesis, Properties and Applications

Dr. Vikas Mittal
Editor

CWP
Central West Publishing

Disclaimer
Every effort has been made by the publisher, editor and authors while preparing this book, however, no warranties are made regarding the accuracy and completeness of the content. The publisher, editor and authors disclaim without any limitation all warranties as well as any implied warranties about sales, along with fitness of the content for a particular purpose. Citation of any website and other information sources does not mean any endorsement from the publisher and authors. For ascertaining the suitability of the contents contained herein for a particular lab or commercial use, consultation with the subject expert is needed. In addition, while using the information and methods contained herein, the practitioners and researchers need to be mindful for their own safety, along with the safety of others, including the professional parties and premises for whom they have professional responsibility. To the fullest extent of law, the publisher, editor and authors are not liable in all circumstances (special, incidental, and consequential) for any injury and/or damage to persons and property, along with any potential loss of profit and other commercial damages due to the use of any methods, products, guidelines, procedures contained in the material herein.

 A catalogue record for this book is available from the National Library of Australia

NATIONAL LIBRARY OF AUSTRALIA

ISBN (print): 978-1-925823-49-3
ISBN (e-book): 978-1-925823-48-6

Contents

3. Characteristics and Applications of Bacterial Nanocellulose   53

Alicia N. Califano, María Laura Balquinta, Patricia Cerrutti, Silvina C. Andrés and Gabriel Lorenzo

4. Nanocellulose as Reinforcement in Polymer Nanocomposites   77

Haleema Saleem and Vikas Mittal

Preface

Cellulose is the most abundant organic compound generated from bi-omass. It has been utilized for around 150 years in numerous applications. Progressing knowledge about the reactivity and structural features of cellulose has led to the gradual development of novel cellulose-based materials. Specifically, the development of nanocellulose in the form of nanocrystals, nano-fibrils, nano-whiskers and nanofibers has gained significant research attention. Novel techniques for the generation of these nanomaterials range from top-down strategies, including physical/synthetic/enzymatic approaches for detachment from agricultural/forest residues and wood, to the bottom-up generation from glucose by bacteria. These materials display unique properties owing to their nanostructure such as hydrophilicity, ease of chemical modification and the generation of adaptable semi-crystalline fiber structures, along with extensive surface area. These properties enable the use of such nanomaterials for a large number of advanced applications. In this context, the book provides deep insights into the current state of the art as well as future trends with respect to the synthetic procedures, property development and applications of nanocellulose.

Chapter 1 reviews the preparation and properties of nanocellulose, along with focusing on various industrial and medical applications. Chapter 2 highlights the suitability of nanocellulose-based materials for removing different kinds of contaminants from water bodies. In addition, future perspective on the use of nanocellulose adsorbents for water remediation is also presented in the chapter. Chapter 3 specifically reviews the characteristics and applications of bacterial nanocellulose. Chapter 4 focuses on the potential application of nanocellulose as reinforcing agent for the generation of polymer nanocomposites. Chapter 5 discusses the use of rice hulls as a raw material to obtain cellulose and nanocellulose, along with focusing on the challenges and advantages of using this lignocellulosic residue as well as the protocols reported in the literature studies for this purpose. Chapter 6 sheds light on different aspects of nanocellulose materials, particularly nano-crystalline celluloses, which make them sustainable materials of choice for 21st century.

The book would not have been successfully accomplished without the support of chapter contributors. The book is dedicated to my family for unswerving support, constant motivation as well as constructive suggestions for improvement.

Vikas MITTAL

1

Nanocellulose: A Review of Preparation, Properties and Applications

Suman,* Nitesh Kumar and Abhishek Kardam

Amity Institute for Advanced Research and Studies (Materials & Devices), Amity University, Noida-201303, U.P., India

Corresponding author: snagpal@amity.edu

1.1 Introduction: Origin and Evolution of Nanocellulose

Nanocellulose was first produced in 1983 by Herrick and Turbak (Figure 1.1). The terminology microfibrillated/nanocellulose or (MFC) was first used by Turbak, Snyder and Sandberg in the late 1970s at the ITT Rayonier labs in Whippany, New Jersey to describe a product prepared as a gel type material by passing wood pulp through a Gaulin type milk homogenizer at high temperatures and high pressures followed by impact ejection against a hard surface. Thereafter, a large number of nanocellulose based materials and composites came into existence with time and this evolution is still in progress.

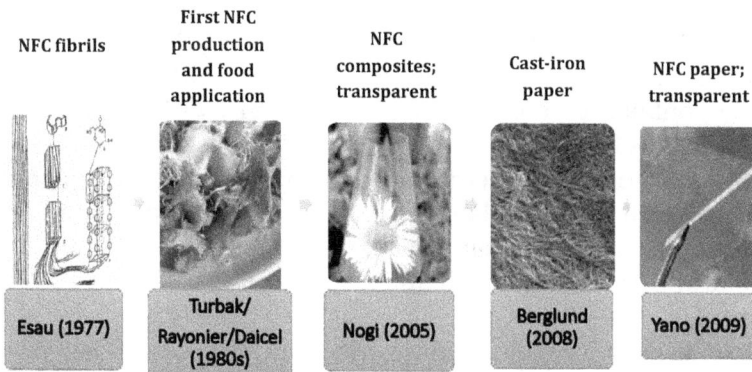

Figure 1.1 History and evolution of nanocellulose.

Nanocellulose, edited by Vikas Mittal
© 2019 Central West Publishing, Australia

1.2 Classification of Nanocellulose

Nanocellulose is classified as

- Cellulose nanocrystals (CNC) – nanocrystalline cellulose, cellulose whiskers, nanowhiskers and nanorods, etc.
- Cellulose nano-fibrils – nano-fibrillated cellulose, micro fibrillated cellulose (MFC) and cellulose nanofibers
- Bacterial cellulose

Cellulose nanocrystals refer to short and rigid rods; cellulose nano-fibril refers to long thread-like cellulose nanofibers and bacterial cellulose refers to ribbon-like bacterial nanocellulose (BNC). Cellulose nanocrystals and cellulose nano-fibrils can be extracted from various sources such as wood, cotton, hemp, flax, wheat straw, sugar beet, potato tuber, mulberry bark, ramie, algae and tunicin.

1.3 Preparation Methods of Nanocellulose from Cellulose

Nanocellulose is extracted from natural cellulose. Due to hierarchical structure and semi-crystalline nature of cellulose, nanoparticles can be extracted by either top-down mechanical process or chemical induced deconstruction strategy by converting the large units (cm) into the smaller ones (nm).

Cellulose (Figure 1.2) is the most abundant organic polymer with formula $(C_6H_{10}O_5)_n$ and is widespread in higher plants, marine algae, bacteria and other biomass [1]. It is a polysaccharide consisting of a linear chain of several hundred to many thousands of D-glucopyranose units linked by $\beta(1 \rightarrow 4)$ glycosidic bonds [2,3]. It mainly contains carbon (44.44%), hydrogen (6.17%) and oxygen (49.39%). It is insoluble in water as well as dilute acidic and alkaline solutions at normal temperatures. Cellulose is the main component of the cell wall of green plants, many forms of algae and oomycetes. Some species of bacteria secrete it to form biofilms [4]. The cellulose content in cotton fiber, wood and dried hemp is 90%, 40-50% and approx. 57% respectively [5].

Various factors, such as maturity, separation processes, microscopic and molecular defects such as pits and nods, type of soil and weather conditions under which they are grown, affect the fiber properties. Macro- or micro-structure of cellulose consists of amorphous as well as crystalline phases. It is relatively easy to break the amorphous phase of cellulosic biopolymer by mechanical, chemical or enzymatic means.

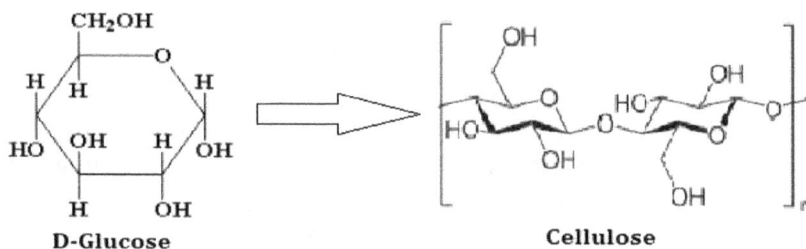

Figure 1.2 Structure of glucose and cellulose.

Extraction of nanocellulose from the cellulosic biomass includes two major steps, viz. pretreatment and removal of amorphous phase by appropriate methods [6]. Prior to the mechanical, chemical or enzymatic treatment, pretreatments such as alkali treatment and bleaching are required.

1.3.1 Mechanical Method

Nanocellulose can be obtained by disintegrating cellulose from the lignocellulosic biomass by mechanical means [7]. Cherian *et al.* [8] reported acid coupled steam treatment method for the preparation of nanocellulose from pineapple leaf fiber. Prior to the mechanical treatment, pretreatments such as grinding, acid hydrolysis, de-crystallization and derivatization are used for the preparation of nanocellulosic materials. Nanocellulose can be derived from the lignocellulosic biomass by high pressure homogenization. The biomass could be suspended by high speed stirring coupled with ultrasonication treatment prior to the high pressure homogenization [9]. Ionic liquid can be used to treat the lignocellulosic biomass prior to the mechanical treatment. The ionic liquid permeates through the microstructure of cellulose and subsequently attacks the hydrogen bonds between cellulose molecules. During high pressure homogenization, inter- and intramolecular bonds are further destroyed, hence, nanocellulose is disintegrated from the biomass.

During mechanical method, multiple mechanical shearing action is applied to the cellulosic fibers to delaminate the fiber effectively into fewer individual micro-fibrils. The product obtained by mechanical method is not a single fiber, so it is called as nano-fibrils

or micro-fibrillated cellulose (MFC). Pre-treatment can also be performed to introduce the electrostatic repulsion between the fibers
and prevent the aggregation.

1.3.2 Chemical Methods

Acid hydrolysis is a common method to extract nanocellulose from
the lignocellulosic biomass [10-13]. After hemicellulose removal and
prior to the acid hydrolysis treatment, the lignocellulosic biomass
can be chemically treated, such as with dimethyl sulfoxide, to swell
the biomass matrix. This helps the acid to diffuse into the domain
structure of lignocellulosic biomass easily and disintegrate the
nanowhiskers [13]. Preparation of nanocellulose from the lignocellulosic biomass by using catalysts can be considered a green approach. Ionic liquid based catalyst has several advantages, viz. wide
range of electrochemical stability, good electrical conductivity, high
ionic mobility and selective dissolution properties towards many
organic and inorganic substances as well as excellent chemical and
thermal stabilities [14]. Ultrasonication has significant influence on
the properties of nanocellulose prepared from bleached hardwood
kraft pulp oxidized by 2,2,6,6-tetramethylpiperidine-1-oxyl (TEM-
PO). Nanocellulose prepared by using ultrasound assisted process is
thinner as well as contains more carboxylic functionality, higher degree of fibrillation and higher yield [15,16].

   For chemical de-structuring process using acid hydrolysis, a
strong acid is utilized for the extraction of crystalline nanocellulose
from natural cellulose by hydrolysis of the cellulose fibers, solubilization and removal of amorphous regions and preservation of the
highly crystalline structure by longitudinal cutting of the microfibrils.

1.3.3 Bacterial Method

Bacterial cellulose is (BC) typically synthesized by bacteria (such as
Acetobacter xylinum) in a pure form which does not require intensive processing to remove unwanted impurities or contaminants
such as lignin, pectin and hemicelluloses. During the biosynthesis of
BC, the glucose chains are produced inside the bacterial body and
extruded out through tiny pores present on the cell envelope. With
the combination of glucose chains, micro-fibrils are formed, which
further aggregate as ribbons (nanofibers).

Nanocellulose obtained by using bacterial method has similar chemical structure as nanocellulose extracted from the lignocellulosic biomass by following chemical and mechanical methods. In addition, ultrafine nanofiber network formed by an appropriate culture medium exhibits unique properties, e.g. high purity, uniform morphology, good water absorption capacity, excellent mechanical properties and flexibility [17-19].

1.4 Unique Properties of Nanocellulose

1.4.1 Physico-chemical Properties

Chemically, both cellulose and nanocellulose consist of β-anhydroglucose units. These units are linked together by β-1-4 glycosidic bonds. Each anhydroglucose unit has two secondary hydroxyl groups at 2, 3 positions and primary hydroxyl group at 6th position. Overall, nanocellulose contain large number of hydroxyl groups (Figure 1.3).

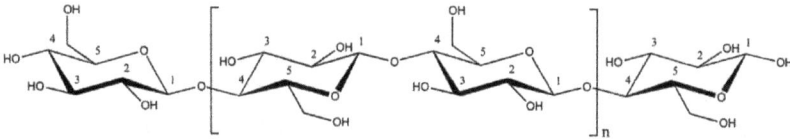

Figure 1.3 Structure of nanocellulose.

1.4.2 Physical Properties

Physically, both cellulose and nanocellulose are white in color, non-toxic, biodegradable. These are solid homopolymers, insoluble in water and soluble in organic chemicals, with high tensile and compressive strength. Nanocellulose materials have other unique properties compared to its bulk raw material (cellulose). Nanocellulose provides extremely larger surface area to volume ratio, which results in significant surface dependent material properties, morphology and spatial confinement.

1.4.3 Morphological Properties

Morphological structure of nanocellulose is highly dependent on the efficient removal of non-cellulosic part and dissolution of amorphous region of domain structure [8,10]. Morphology, physical properties and dimensions of nanocellulose vary significantly with different natural fiber resources and extraction processes. The morphology of nanocellulose-based materials extracted by acid hydrolysis from sisal, pineapple leaf and coir exhibit long, flexible and entangled morphology of nanofibers, whereas more individualized and rod like structures are visible when nanocelluloses are extracted from the banana rachis and kapok [20].

1.4.4 Mechanical Properties

Mechanical properties of the nanocellulose are based on the crystalline (ordered) and amorphous (disordered) regions. In nanocellulose, the amorphous region is responsible for the flexibility and plasticity, whereas the crystalline region is responsible for the stiffness and elasticity of the material. The crystallinity of nanocellulose is determined by wave propagation, X-ray diffraction, Raman spectroscopy and atomic force microscopy.

In comparison with their lignocellulosic source material, nanocellulose has superior mechanical properties due to more uniform morphological structure. The average modulus of nanocellulose is 100 gigapascals (GPa), which is much higher than the base cellulosic materials [21]. Crystal structure and degree of crystallinity of nanocellulose depend on the lignocellulosic sources [22]. Nanocellulose contains both cellulose I and cellulose II, which is characterized by typical X-ray diffraction peaks at around 2θ ~23° and ~34° respectively [23].

1.5 Applications of Nanocellulose

Superior properties of nanocellulose make it an interesting material for a variety of applications. A few applications are as follows:

1.5.1 Industrial Applications

Paper Industry

There is a potential of nanocellulose applications in the area of paper and paperboard manufacturing. Nanocellulose-based materials are expected to enhance the fiber-fiber bond strength, hence, resulting in strong reinforcement effect on paper materials [24]. Nanocellulose may also be useful as a barrier in grease-proof paper and as a wet-end additive to enhance retention, dry and wet strength in commodity paper and board products [25]. It has been shown that application of nanofibers as a coating on the surface of paper and paperboard improves the barrier properties, especially air resistance [26]. It also enhances the structure properties of paperboards (smoother surface). In addition, nanocellulose can be used to prepare flexible and optically transparent paper. Such paper is an attractive substrate for electronic devices as it is recyclable, compatible with biological objects and easily degrades when disposed.

Emulsion and Dispersion

Nanocellulose has various emulsion and dispersion applications in different fields [27]. Oil in water applications was recognized much earlier. Early investigations explored non-settling suspensions for pumping sand, coal, paints and drilling muds.

Oil Recovery

Hydrocarbon fracturing of oil-bearing formations is a potentially beneficial and large-scale application. Nanocellulose has been suggested for use in oil recovery applications as a fracturing fluid. Drilling muds based on nanocellulose have also been suggested.

1.5.2 Medical Applications

A wide range of applications of nanocellulose in medical and pharmaceutical industries has been reported, as mentioned in Figure 1.4.

Life Cycle and Biological Impact of Nanocellulose

It is necessary to evaluate the potential risk of nanocellulose for human health by studying its life cycle. According to Shatkin and Kim [29], life cycle of nanocellulose-based composite materials is categorized in 5 different stages which can be identified as: Stage 1 (production or isolation of raw materials), Stage 2 (manufacturing of products), Stage 3 (transportation of products), Stage 4 (consumer use of products) and Stage 5 (disposal of products), as shown in Figure 1.5 [29]. In the life-cycle risk assessment framework (NANO

LCRA), different exposure scenarios were evaluated and ranked as a function of the potential, magnitude, likelihood and frequency of the

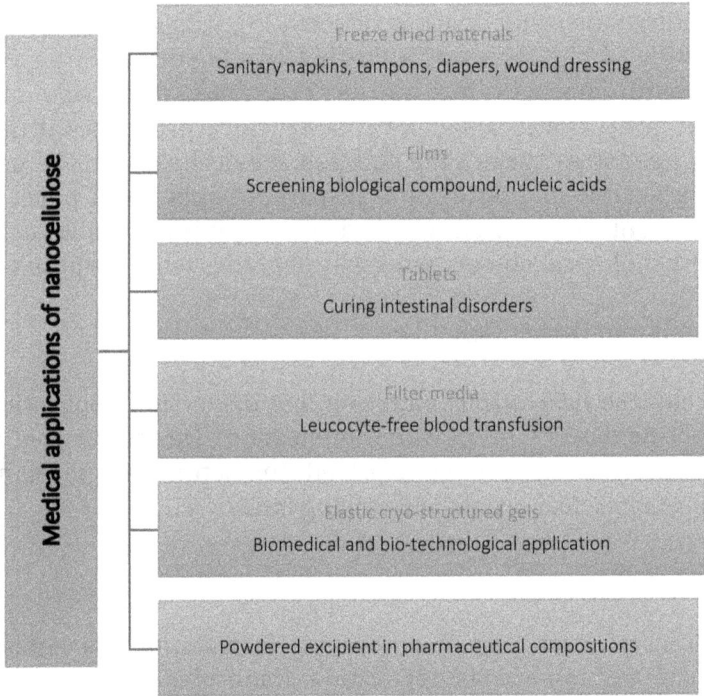

Figure 1.4 Medical application of nanocellulose [28].

hazard. The authors identified the top four exposure scenarios to be (1) inhalation of dry raw material by a facility employee during production, (2) application of dry raw nanocellulose to create a film and inhalation during manufacturing, (3) inhalation of dry raw nanocellulose powder during mixing with other materials to manufacture a product, and (4) inhalation by incidental contact with the raw form of nanocellulose. In this experiment, transportation was not considered during the evaluation of the life cycle. Analysis of the data suggested that the main exposure route was the inhalation of (raw) nanocellulose [29,30]. Polymer nanocomposites might also have led to the inhalation of cellulose/polymer particles during processing such as drilling, cutting, sanding, etc. [31]. For many applications, such as healthcare products, cellulose might be surface functionalized to impart new properties to the material, thus, triggering the need of an independent case study for such materials [32,33].

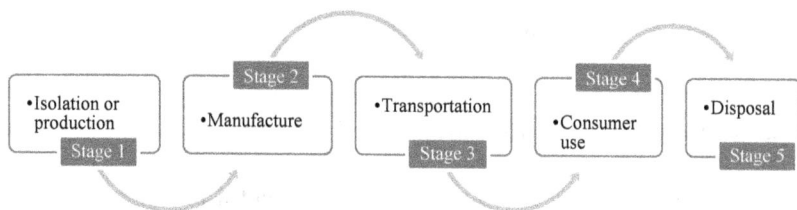

Figure 1.5 Life cycle of nanocellulose-based material.

For biological impact of nanocellulose, main biochemical endpoints have been studied, including cytotoxicity, inflammatory response, oxidative stress and genotoxicity. Throughout the particle and fiber toxicology field, these endpoints are recognized as the most important drivers of nanomaterial toxicity [34].

Cytotoxicity

Several studies have confirmed the limited toxic potential of nanocellulose in terms of cytotoxicity, using various experimental systems on cellulose-human interactions. A sophisticated triple-cell co-culture model of the human epithelial tissue barrier was used in one of the studies, which showed no significant cytotoxicity of two different cellulose nanocrystal types isolated from cotton and tunicates (which were deposited onto the cells in different doses from aerosolized water-based suspensions). However, clearance, albeit based upon the dose, time and CNC-dependent manner, of deposited CNCs by macrophages was observed, with a lower efficiency associated with the tunicate CNC [30].

The incidence of benign results in terms of cytotoxicity, viability and impact of nanocellulose on mammalian cell morphology seems to be prevalent in the current literature. The adverse effects related to nanocellulose exposure have to be taken into consideration when evaluating the overall hazard posed by this material. Many observed differences can be attributed to the variation in cell systems, material origin, treatment and characterization, cell exposure doses reaching non-realistic concentrations of nanocellulose, exposure scenarios, biological systems, etc. Due to the nature of nanocellulose, it is challenging to track it due to a lack of analytical methods feasible to measure nanocellulose in biological systems. Therefore, the information about morphological impact or organ distribution of nanocellulose after exposure is limited. Nevertheless, the overall results can be interpreted that most of the studies hint at a limited hazard potential of nanocellulose.

1.5.3 Nanocellulose Composites

Nanocellulose has also been reported as a useful material for reinforcing plastics to improve the mechanical properties of thermosetting resins, starch-based matrices, soy protein, rubber latex, poly(lactide), etc. The composites are useful for applications such as coatings and films, paints, foams, packaging, among others. Nanocellulose composites have also been reported as bio-adsorbent for heavy metals, dyes, dissolved organic pollutants, oil and undesired effluents.

1.5.4 Other Applications

Figure 1.6 shows other application of nanocellulose.

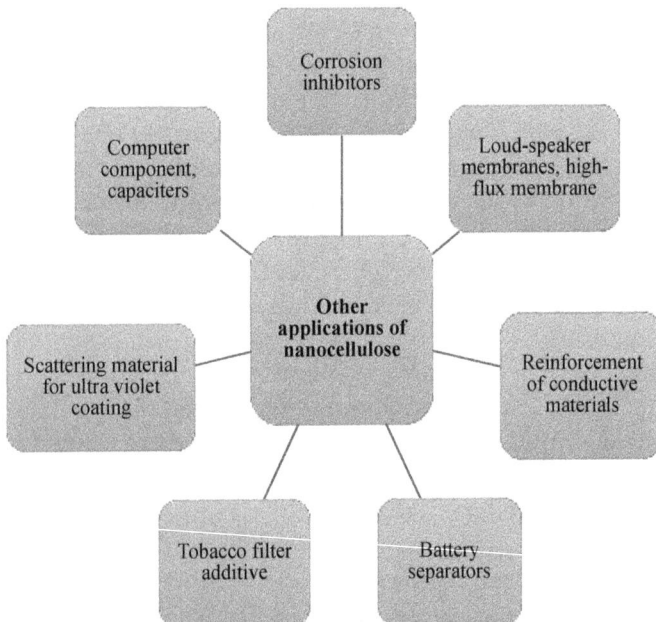

Figure 1.6 Other applications of nanocellulose.

Figure 1.7 also shows the statistical representation of application of nanocellulose based on market segments.

2011

Rheological modifiers 5%

Electronics 4%

Paper and board 21%

NC Composites 46%

Coatings 7%

Filtration 8%

Medical, cosmetic & pharmaceutical 6%

Aerogels 3%

2016

Rheological modifiers 6%

Electronics 8%

Paper and board 20%

NC Composites 36%

Coatings 5%

Filtration 10%

Aerogels 4%

Medical, cosmetic & pharmaceutical 11%

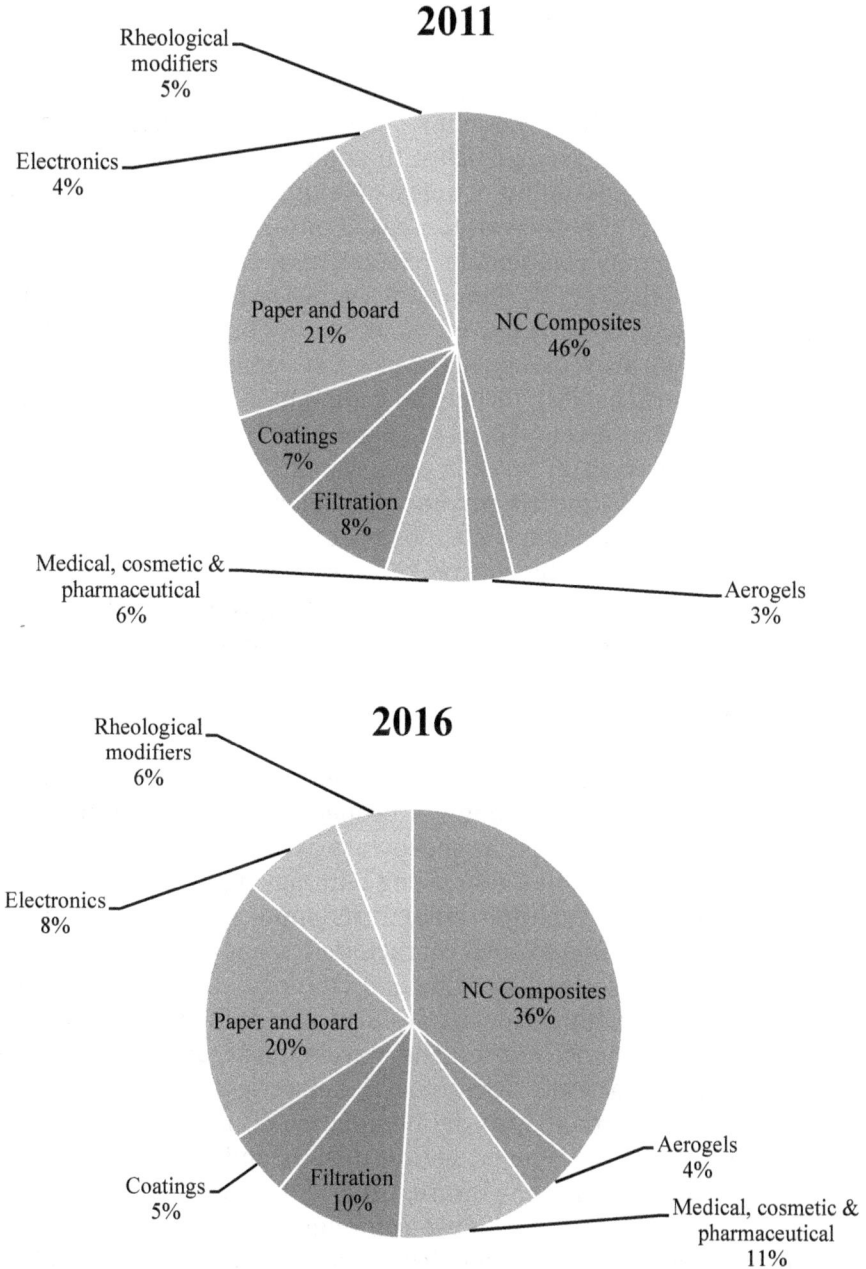

Figure 1.7 Statistical representation of application of nanocellulose based on market segments. Source: Nanocellulose Market Study, Future Markets Inc, 2012.

1.5.5 Nanocellulose as Adsorbent for Heavy Metals

Nanocellulose is a lightweight material with strong mechanical strength, inexpensive production costs and safe handling, as compared to synthetic nanoparticles. In addition, high specific surface area and broad possibility of surface modification contribute to its emergence as a new class of bio-based adsorbents with promising applications in environmental remediation. A large variety of pollutants can be adsorbed by nanocellulose, including dissolved organic pollutants, dyes, oils and undesired effluents. Some of authors' earlier studies have also confirmed the effective removal of heavy metals (Table 1.1) such as lead, nickel and cadmium [35]. Recently, the authors have also developed novel nanocellulose composites based columns, providing a three-in-one system, which effectively removes dyes, heavy metals and microbial loads from waste water. In addition, the nanocellulose composites can be reused up to five cycles [36]. The possibility of the regeneration of the nanocellulose adsorbent is another benefit, which is driving research attempts to fully exploit this new class of nanostructured bio-based material.

1.6 Conclusions

Nanocellulose-based materials represent a class of novel nanomaterials having a wide range of properties, both in the modified and non-modified forms. Nanocellulose is natural, biodegradable, biocompatible and renewable, along with having high surface area, large aspect ratios, superior strength and modulus, ease of chemical modification, etc. Due to such characteristics, nanocellulose is widely used as bio-adsorbent in dying and tanning industries. The fields of coatings and medical devices are also extremely benefited by the application of nanocellulose. The development of nanocellulose-based organic materials with tunable, 'smart' and biomimetic characteristics will further push the boundaries of nanocellulose application in flexible electronics, optical devices and high performance functional plastics in near future, especially as cost-effective commercial sources of nanocellulose continue to emerge. Overall, even though the applications of nanocellulose may be somewhat limited by availability and cost, the outlook is, however, promising, as more and more industrial entities and researchers engage their efforts for the development of solutions to many existing challenges through the use of nanocellulose.

Table 1.1 Nanocellulose as adsorbent for heavy metals

Sr. no.	Adsorbents	Modifying agent	Heavy metals (Adsorbate)	Ref.
1	Nanocellulose	Enzymatically phosphorylated	Ag^+, Cu^{2+} and Fe^{3+}	[37]
2	Cellulose nanofibers	2,2,6,6-tetramethyl-1-piperidinyloxy (TEMPO)	Cu^{2+}, Ni^{2+}, Zn^{2+} and Cr^{3+}	[38]
3	Carboxylated cellulose nanofibrils	Magnetic chitosan hydrogel beads	Pb^{2+}	[39]
4	Wheat pulp nanocelluloses	Sulfonated	Pb^{2+}	[40]
5	Nanocellulose	Polyhedral oligomeric silsesquioxanes	Cu^{2+} and Ni^{2+}	[41]
6	Microfibrillated cellulose (MFC)	Aminopropyltriethoxysilane (APS)	Ni^{2+}, Cu^{2+} and Cd^{2+}	[42]
7	Nanocellulose fibers	Sulfuric acid (carboxyl)	Cd^{2+}, Pb^{2+} and Ni^{2+}	[35]
8	Nanofibrillated cellulose	Poly(methacrylic acid-co-maleic acid)	Pb^{2+}, Cd^{2+}, Zn^{2+} and Ni^{2+}	[43]
9	Cellulose nanofiber	Quaternary ammonium-functionalized	Cr^{6+}	[44]
10	Nanocelluloses	-	-	[45]
11	Nanocellulose	Silver nanoparticles (AgNPs) embedded pebbles	Pb^{2+}, Cr^{3+}	[36]
12	Nanocellulose	-	Pb^{2+}	[46]
13	Nanocellolose fiber	-	Cd^{2+}	[47]
14	Cellulose nanocomposite membranes	cellulose microfiber sludge and cellulose nanocrystals	Ag^+ and $Cu^{2+}/Fe^{3+}/Fe^{2+}$	[48]
15	Nanocellulose/nanobentonite	Anchored with multi-carboxyl functional groups	Co^{2+}	[49]
16	Nanocellulose	Succinic anhydride (Carboxyl)	Zn^{2+}, Cd^{2+}, Cu^{2+}, Co^{2+}. Ni^{2+}	[50]
17	Cellulose nanofibrils	Phosphorylated	Cu^{2+}	[51]
18	Cellulose nanocrystals	Magnetic carboxylated	Pb^{2+}	[52]
19	Cellulose nanofibrils	Aldehyde functionalized	Cu^{2+} and Pb^{2+}	[53]

References

1. Klemm, D., Heublein, B., Fink, H. P., and Bohn, A. (2005) Cellulose:

fascinating biopolymer and sustainable raw material. *Angewandte Chemie International Edition.* **44**(22), 3358-3393.

2. Crawford, R. L. (1981) *Lignin Biodegradation and Transformation,* John Wiley and Sons, USA.
3. Updegraff, D. M. (1969) Semimicro determination of cellulose in biological materials. *Analytical Biochemistry,* **32**(3), 420-424.
4. Otto, M. (2008) Staphylococcal biofilms. In: *Bacterial Biofilms,* Tony, R. (ed.), Springer, Germany, pp. 207-228.
5. Piotrowski, S., and Carus, M. (2011) *Multi-criteria Evaluation of Lignocellulosic Niche Crops for Use in Biorefinery Processes,* Nova-Institut GmbH, Germany. Online: http://www.biocore-europe.org/file/BIOCORE%20Multi-criteria%20evaluation%20of%20niche%20crops.pdf [accessed 17th January 2019].
6. Kim, J.-H., Shim, B. S., Kim, H. S., Lee, Y.-J., Min, S.-K., Jang, D., Abas, Z., and Kim, J. (2015) Review of nanocellulose for sustainable future materials. *International Journal of Precision Engineering and Manufacturing - Green Technology,* **2**(2), 197-213.
7. Hettrich, K., Pinnow, M., Volkert, B., Passauer, L., and Fischer, S. (2014) Novel aspects of nanocellulose. *Cellulose,* **21**(4), 2479-2488.
8. Cherian, B. M., Leão, A. L., de Souza, S. F., Thomas, S., Pothan, L. A., and Kottaisamy, M. (2010) Isolation of nanocellulose from pineapple leaf fibres by steam explosion. *Carbohydrate Polymers,* **81**(3), 720-725.
9. Wang, H., Zhang, X., Jiang, Z., Li, W., and Yu, Y. (2015) A comparison study on the preparation of nanocellulose fibrils from fibers and parenchymal cells in bamboo (Phyllostachys pubescens). *Industrial Crops and Products,* **71**, 80-88.
10. Lu, Z., Fan, L., Zheng, H., Lu, Q., Liao, Y., and Huang, B. (2013) Preparation, characterization and optimization of nanocellulose whiskers by simultaneously ultrasonic wave and microwave assisted. *Bioresource Technology,* **146**, 82-88.
11. Lin, N., Bruzzese, C. C., and Dufresne, A. (2012) TEMPO-oxidized nanocellulose participating as crosslinking aid for alginate-based sponges. *ACS Applied Materials & Interfaces,* **4**(9), 4948-4959.
12. Morais, J. P. S., de Freitas Rosa, M., Nascimento, L. D., do Nascimento, D. M., and Cassales, A. R. (2013) Extraction and characterization of nanocellulose structures from raw cotton linter. *Carbohydrate Polymers,* **91**(1), 229-235.
13. Mandal, A., and Chakrabarty, D. (2011) Isolation of nanocellulose from waste sugarcane bagasse (SCB) and its characterization. *Carbohydrate Polymers,* **86**(3), 1291-1299.
14. Hamid, S. B. A., Amin, M., and Ali, M. E. (2014) Zeolite supported ionic liquid catalyst for the synthesis of nano-cellulose from palm tree biomass. *Advanced Materials Research,* **925**, 52-56.

15. Mishra, S. P., Manent, A.-S., Chabot, B., and Daneault, C. (2012) The use of sodium chlorite in post-oxidation of TEMPO-oxidized pulp: Effect on pulp characteristics and nanocellulose yield. *Journal of Wood Chemistry and Technology*, **32**(2), 137-148.

16. Mishra, S. P., Thirree, J., Manent, A.-S., Chabot, B., and Daneault, C. (2010) Ultrasound-catalyzed TEMPO-mediated oxidation of native cellulose for the production of nanocellulose: Effect of process variables. *BioResources*, **6**(1), 121-143.

17. Taokaew, S., Seetabhawang, S., Siripong, P., and Phisalaphong, M. (2013) Biosynthesis and characterization of nanocellulose-gelatin films. *Materials*, **6**(3), 782-794.

18. Iguchi, M., Yamanaka, S., and Budhiono, A. (2000) Bacterial cellulose - A masterpiece of nature's arts. *Journal of Materials Science*, **35**(2), 261-270.

19. Chen, S., Zou, Y., Yan, Z., Shen, W., Shi, S., Zhang, X., and Wang, H. (2009) Carboxymethylated-bacterial cellulose for copper and lead ion removal. *Journal of Hazardous Materials*, **161**(2-3), 1355-1359.

20. Deepa, B., Abraham, E., Cordeiro, N., Mozetic, M., Mathew, A. P., Oksman, K., Faria, M., Thomas, S., and Pothan, L. A. (2015) Utilization of various lignocellulosic biomass for the production of nanocellulose: a comparative study. *Cellulose*, **22**(2), 1075-1090.

21. Dufresne, A. (2013) Nanocellulose: A new ageless bionanomaterial. *Materials Today*, **16**(6), 220-227.

22. Le Bras, D., Strømme, M., and Mihranyan, A. (2015) Characterization of dielectric properties of nanocellulose from wood and algae for electrical insulator applications. *The Journal of Physical Chemistry B*, **119**(18), 5911-5917.

23. Peng, Y., Gardner, D. J., Han, Y., Kiziltas, A., Cai, Z., and Tshabalala, M. A. (2013) Influence of drying method on the material properties of nanocellulose I: thermostability and crystallinity. *Cellulose*, **20**(5), 2379-2392.

24. Taipale, T., Österberg, M., Nykänen, A., Ruokolainen, J., and Laine, J. (2010) Effect of microfibrillated cellulose and fines on the drainage of kraft pulp suspension and paper strength. *Cellulose*, **17**(5), 1005-1020.

25. Syverud, K., and Stenius, P. (2009) Strength and barrier properties of MFC films. *Cellulose*, **16**(1), 75.

26. Jung, Y. H., Chang, T.-H., Zhang, H., Yao, C., Zheng, Q., Yang, V. W., Mi, H., Kim, M., Cho, S. J., Park, D.-W., Jiang, H., Lee, J., Qiu, Y., Zhou, W., Cai, Z., Gong, S., and Ma, Z. (2015) High-performance green flexible electronics based on biodegradable cellulose nanofibril paper. *Nature Communications*, **6**, 7170.

27. Xhanari, K., Syverud, K., and Stenius, P. (2011) Emulsions stabilized by microfibrillated cellulose: the effect of hydrophobization, concentration and o/w ratio. *Journal of Dispersion Science and*

Technology, **32**(3), 447-452.

28. Syverud, K., Kirsebom, H., Hajizadeh, S., and Chinga-Carrasco, G. (2011) Cross-linking cellulose nanofibrils for potential elastic cryo-structured gels. *Nanoscale Research Letters*, **6**(1), 626.

29. Shatkin, J. A. and Kim, B. (2015) Cellulose nanomaterials: life cycle risk assessment, and environmental health and safety roadmap. *Environmental Science: Nano*, **2**(5), 477-499.

30. Endes, C., Camarero-Espinosa, S., Mueller, S., Foster, E., Petri-Fink, A., Rothen-Rutishauser, B., Weder, C., and Clift, M. (2016) A critical review of the current knowledge regarding the biological impact of nanocellulose. *Journal of Nanobiotechnology*, **14**(1), 78.

31. Köhler, A. R., Som, C., Helland, A., and Gottschalk, F. (2008) Studying the potential release of carbon nanotubes throughout the application life cycle. *Journal of Cleaner Production*, **16**(8-9), 927-937.

32. Albanese, A., Tang, P. S., and Chan, W. C. (2012) The effect of nanoparticle size, shape, and surface chemistry on biological systems. *Annual Review of Biomedical Engineering*, **14**, 1-16.

33. Clift, M. J., Rothen-Rutishauser, B., Brown, D. M., Duffin, R., Donaldson, K., Proudfoot, L., Guy, K., and Stone, V. (2008) The impact of different nanoparticle surface chemistry and size on uptake and toxicity in a murine macrophage cell line. *Toxicology and Applied Pharmacology*, **232**(3), 418-427.

34. Stone, V., Miller, M. R., Clift, M. J., Elder, A., Mills, N. L., Møller, P., Schins, R. P. F., Vogel, U., Kreyling, W. G., Jensen, K. A., Kuhlbusch, T. A. J., Schwarze, P. E., Hoet, P., Pietroiusti, A., De Vizcaya-Ruiz, A., Baeza-Squiban, A., Teixeira, J. P., Tran, C. L., and Cassee, F. R. (2017) Nanomaterials versus ambient ultrafine particles: an opportunity to exchange toxicology knowledge. *Environmental Health Perspectives*, **125**(10), 106002.

35. Kardam, A., Raj, K. R., Srivastava, S., and Srivastava, M. (2014) Nanocellulose fibers for biosorption of cadmium, nickel, and lead ions from aqueous solution. *Clean Technologies and Environmental Policy*, **16**(2), 385-393.

36. Suman, Kardam, A., Gera, M., and Jain, V. (2015) A novel reusable nanocomposite for complete removal of dyes, heavy metals and microbial load from water based on nanocellulose and silver nano-embedded pebbles. *Environmental Technology*, **36**(6), 706-714.

37. Liu, P., Borrell, P. F., Božič, M., Kokol, V., Oksman, K., and Mathew, A. P. (2015) Nanocelluloses and their phosphorylated derivatives for selective adsorption of Ag^+, Cu^{2+} and Fe^{3+} from industrial effluents. *Journal of Hazardous Materials*, **294**, 177-185.

38. Sehaqui, H., de Larraya, U. P., Liu, P., Pfenninger, N., Mathew, A. P., Zimmermann, T., and Tingaut, P. (2014) Enhancing adsorption of heavy metal ions onto biobased nanofibers from waste pulp residues for application in wastewater treatment. *Cellulose*, **21**(4),

2831-2844.

39. Zhou, Y., Fu, S., Zhang, L., Zhan, H., and Levit, M. V. (2014) Use of carboxylated cellulose nanofibrils-filled magnetic chitosan hydrogel beads as adsorbents for Pb (II). *Carbohydrate polymers*, **101**, 75-82.

40. Suopajärvi, T., Liimatainen, H., Karjalainen, M., Upola, H., and Niinimäki, J. (2015) Lead adsorption with sulfonated wheat pulp nanocelluloses. *Journal of Water Process Engineering*, **5**, 136-142.

41. Xie, K., Jing, L., Zhao, W., and Zhang, Y. (2011) Adsorption removal of Cu^{2+} and Ni^{2+} from waste water using nano-cellulose hybrids containing reactive polyhedral oligomeric silsesquioxanes. *Journal of Applied Polymer Science*, **122**(5), 2864-2868.

42. Hokkanen, S., Repo, E., Suopajärvi, T., Liimatainen, H., Niinimaa, J., and Sillanpää, M. (2014) Adsorption of Ni (II), Cu (II) and Cd (II) from aqueous solutions by amino modified nanostructured microfibrillated cellulose. *Cellulose*, **21**(3), 1471-1487.

43. Maatar, W., and Boufi, S. (2015) Poly (methacylic acid-co-maleic acid) grafted nanofibrillated cellulose as a reusable novel heavy metal ions adsorbent. *Carbohydrate Polymers*, **126**, 199-207.

44. He, X., Cheng, L., Wang, Y., Zhao, J., Zhang, W., and Lu, C. (2014) Aerogels from quaternary ammonium-functionalized cellulose nanofibers for rapid removal of Cr (VI) from water. *Carbohydrate Polymers*, **111**, 683-687.

45. Dwivedi, A. D., Dubey, S. P., Hokkanen, S., and Sillanpää, M. (2014) Mechanistic investigation on the green recovery of ionic, nanocrystalline, and metallic gold by two anionic nanocelluloses. *Chemical Engineering Journal*, **253**, 316-324.

46. Kardam, A., Raj, K. R., Arora, J. K., and Srivastava, S. (2012) Artificial neural network modeling for biosorption of Pb (II) ions on nanocellulose fibers. *Bionanoscience*, **2**(3), 153-160.

47. Kardam, A., Raj, K. R., Arora, J. K., and Srivastava, S. (2013) Simulation and optimization of artificial neural network modeling for prediction of sorption efficiency of nanocellulose fibers for removal of Cd (II) ions from aqueous system. *Walailak Journal of Science and Technology (WJST)*, **11**(6), 497-508.

48. Karim, Z., Mathew, A. P., Kokol, V., Wei, J., and Grahn, M. (2016) High-flux affinity membranes based on cellulose nanocomposites for removal of heavy metal ions from industrial effluents. *RSC Advances*, **6**(25), 20644-20653.

49. Anirudhan, T., Deepa, J., and Christa, J. (2016) Nanocellulose/nanobentonite composite anchored with multi-carboxyl functional groups as an adsorbent for the effective removal of Cobalt (II) from nuclear industry wastewater samples. *Journal of Colloid and Interface Science*, **467**, 307-320.

50. Hokkanen, S., Repo, E., and Sillanpää, M. (2013) Removal of heavy

metals from aqueous solutions by succinic anhydride modified mercerized nanocellulose. *Chemical Engineering Journal*, **223**, 40-47.

51. Mautner, A., Maples, H., Kobkeatthawin, T., Kokol, V., Karim, Z., Li, K., and Bismarck, A. (2016) Phosphorylated nanocellulose papers for copper adsorption from aqueous solutions. *International Journal of Environmental Science and Technology*, **13**(8), 1861-1872.

52. Lu, J., Jin, R.-N., Liu, C., Wang, Y.-F., and Ouyang, X.-k. (2016) Magnetic carboxylated cellulose nanocrystals as adsorbent for the removal of Pb (II) from aqueous solution. *International Journal of Biological Macromolecules*, **93**, 547-556.

53. Yao, C., Wang, F., Cai, Z., and Wang, X. (2016) Aldehyde-functionalized porous nanocellulose for effective removal of heavy metal ions from aqueous solutions. *RSC Advances*, **6**(95), 92648-92654.

2

Nanocellulose-based Materials for Water Purification

Sanna Hokkanen

Laboratory of Green Chemistry, Lappeenranta University of Technology, LUT Savo Sammonkatu 12, FI-50130 Mikkeli, Finland

sanna.hokkanen@lut.fi

2.1 Introduction

Water is the source of life and regarded as the most essential commodity among natural resources. Water is one of the most widely distributed substance on the planet: about 70% of the earth's surface is covered with water. Approximately 97.5 % of this water is salt water, leaving only 2.5% as fresh water. As Figure 2.1 demonstrates, more than 79% of fresh water is locked up in icebergs and glaciers, and less than 1% is available for humans and animals to survive (Figure 2.1).

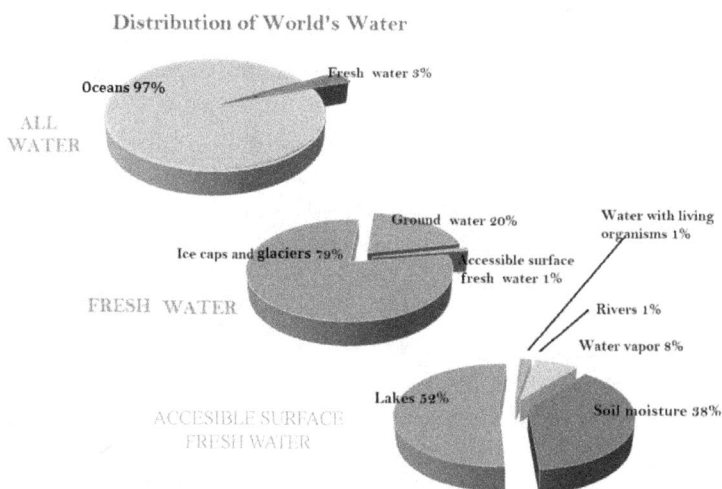

Figure 2.1 Distribution of the world's water.

Nanocellulose, edited by Vikas Mittal
© 2019 Central West Publishing, Australia

Around the world, human activity and natural forces are constantly reducing available water resources. For instance, the increasing worldwide contamination of freshwater systems such as heavy metals, dyes, drugs, pesticides, fluoride, phenols, insecticides, pesticides and detergents is one of the key environmental problems facing humanity [1,2]. The natural disasters such as tsunamis, earthquakes, hurricanes, floods and volcanoes strongly influence water quality. According to World Health Organization, 2.1 billion people are currently without safely managed water services and 159 million people of them collect untreated surface water from lakes, ponds, rivers and streams [3]. It has been predicted that 1.8 billion people will live in countries or regions with absolute water scarcity and two-thirds of the world population could be affected by water stress conditions by 2025 [4]. The decreasing clean water resources, thus, create the need of proper water governance, as the competition for water will increase in future and may cause an escalation of water crises of various kinds. It is well known that fresh water often has a crucial role in managing risks such as famine, migration, epidemics, inequalities and political instability.

The above mentioned challenges have led to a recognition of the need of more effective, lower-cost and robust methods for waste water treatment, without further stressing the environment or endangering human health [5]. Extensive research have been conducted in the recent years with an aim of developing alternative and economically feasible technologies for water and waste water treatment. The use of natural bio-based materials for water treatment has gained wide attention in recent years due to their eco-friendly characteristics, low cost, high uptake capacity, less sludge generation, possible regeneration and abundant availability worldwide [6]. Nanocellulose (NC) is a potential nanomaterial in this context due to the combination of high strength, chemical inertness, hydrophilic surface chemistry and high surface area.

The term nanocellulose includes all of the different analogs of cellulose (cellulose nanofibers (CNF), microfibrillated cellulose (MFC), nanocrystalline cellulose (NCC or CNC) and bacterial nanocellulose). In this review, these analogs are not significantly different from each other. Various nanocellulose-based materials as well as modification methods have been described for water purification, however this chapter focuses on the suitability of nanocellulose-based materials for removing different kinds of contaminants from water bodies. In addition, future perspective on the use of nanocellulose adsorbents for water remediation is also presented in the chapter.

2.2 Adsorption and Membrane Filtration

Adsorption is a spontaneous surface-based process where forces of attraction exist between the adsorbent and adsorbate. The material accumulated at the interface is the adsorbate and the solid surface is the adsorbent. Based on the surface interaction, four types of adsorption occurs: ion exchange, physisorption, chemisorption and specific adsorption. Exchange or ion exchange adsorption encompasses electrostatic attachment of ionic species to the sites of opposite charge at the surface of an adsorbent with subsequent displacement of these species by other ionic adsorbates of greater electrostatic affinity. The characteristic interactions are ion-ion and ion-dipole type [7,8]. Physisorption results from the action of weak van der Waals forces. This type of adsorption has the characteristics of low enthalpy (less than 80 kJ/mol). The adsorption can be monolayer or multilayer, with no dissociation of the adsorbed species and a decrease in adsorption capacity with increasing temperature [8]. On the contrary, chemical adsorption involves strong adsorbate-adsorbent interactions resulting in a change in the chemical form of the adsorbate. Due to its strong interaction, chemisorption has high enthalpy (80-800 kJ/mol), and adsorption occurs at monolayer only, with dissociation of adsorbed species [8]. Specific adsorption results from the interaction between adsorbate and adsorbent which does not result in a chemical change in the adsorbate. The binding energy value of this type of adsorption is in between physisorption and chemisorption [7,8]. Adsorption from a solution onto a solid is dependent on the characteristics of the adsorbent, adsorbate and solution. Either isotherm or kinetic point of view are usually used to understand the nature of the adsorption mechanisms. Three types of adsorption isotherms for a solid-liquid system are presented to describe the adsorption mechanism: type I (favorable), type II (linear) and type III (unfavorable). For favorable adsorption isotherm, adsorption normally occurs on microporous adsorbents where pore size is not greater than the molecular diameter of the adsorbate [8]. Type II shows linear isotherm at low concentration, this type is well known as the classical Langmuir form. Type III is commonly observed for a wide range of adsorbent pore sizes [7]. Adsorption is influenced by the properties of adsorbate-solvent, system and adsorbent. If the adsorbate molecule has high solubility in water, the adsorption is observed to decrease. The system properties such as pH can have a major influence on adsorption mechanism. Adsorbent surface area and distribution with respect to pore size are important to determine adsorption capacity [7,9].

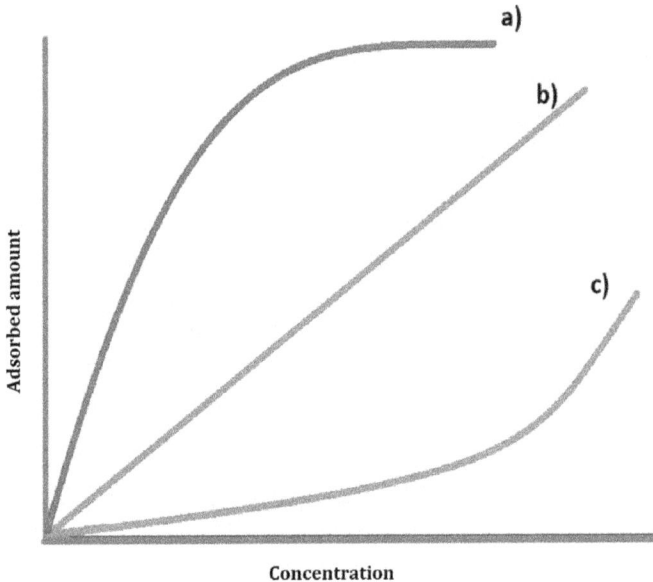

Figure 2.2 Three types of adsorption isotherms, (a) type I (favorable), (b) type II (linear) and (c) type III (unfavorable).

A membrane is a semi-permeable thin layer of material which can separate the contaminants as a function of their size and chemical characteristics. Overall, there are four main types of membrane separation: microfiltration, ultrafiltration, nanofiltration and reverse osmosis (Figure 2.3). The choice of membrane and, thus, the pore size depends on the filtration purpose, which can be disinfection, desalination, organic removal or softening of water and waste water. The membrane types can be divided into two subcategories: 1) low pressure operations including microfiltration (0.1 μm pores) and ultrafiltration (0.01 μm pores); 2) high pressure operations including nanofiltration (0.001 μm pores) and reverse osmosis (<0.001 μm).

2.3 Nanocellulose as Adsorbent and Membrane Material

In recent years, bio-based nanosized particles have been the subject of growing attention in the field of research and product development

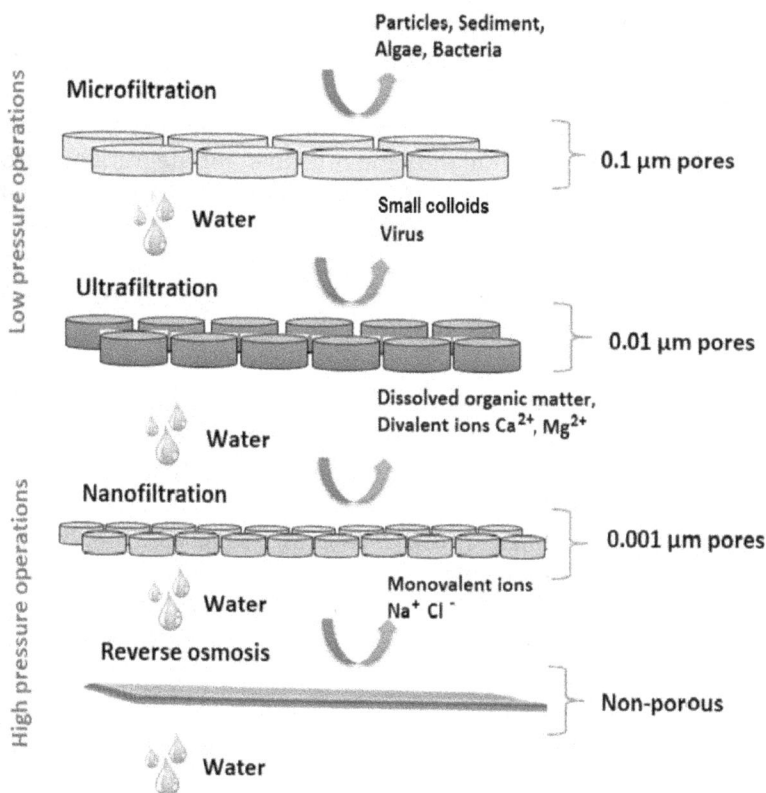

Figure 2.3 Membrane filtration systems.

because of favorable features like non-hazardous character, biodegradability and surface reactivity. Nanocellulose is the most studied bio-based nanomaterial mainly because of the universal availability of cellulose. Nanocellulose can be derived from a wide variety of sources. Wood is considered as the most important cellulose-containing material, however, other materials including agriculture residues, water plants, grasses and other plant products can be used as raw materials for nanocellulose. Agricultural residues have attracted significant interest in recent years as an alternative raw material of nanocellulose due to their abundance all over the world and renewability, especially for countries that do not have wood based raw materials due to the lack of forests. It is noteworthy that the production of nanocellulose has reached the industrial scale in many countries. Many literature studies have confirmed the merits of nanocellulose as an adsorbent for water treatment, which include high surface area and aspect ratio [10,11], high mechanical stiffness [11-13], broad possibility of surface functionalization [2,6,14,15] and stability in water [11,16].

Concerns of environmental contamination, increasing environmental awareness and demands for reusable and degradable products have driven the replacement of non-degradable and toxic materials containing traditional membranes with more effective and environmental friendly products. Natural fibers are ideally suited for this purpose due to their low cost and biodegradability. Thus, fully biodegradable membranes from biofibers and biodegradable polymers have been produced in recent years. Bio-based membranes offer solutions to waste disposal problems associated with traditional petroleum-derived materials. The stability in water environment, hydrophilicity, high surface area, good mechanical strength and rigidity are the required properties of membranes for water treatment purpose. Nanocellulose offers these properties, thus, its use as a film material for water purification purposes has seen recent growth. The presence of large number of functionalities on surface of nanocellulose also enables its use as a unique platform for significant surface modification through various methods.

2.4 Cellulose Modification for Water Treatment Applications

Owing to the molecular structure of cellulose, it is an active material due to the presence of three hydroxyl groups in each anhydroglucose unit. These hydroxyl groups may be partially or completely reacted with various chemicals, thus, resulting in derivatives whose beneficial properties form the basis for water treatment technology and research.

2.4.1 Chemical Modification

The modification of cellulose can be divided into two approached: direct modification and monomer grafting. The direct modification can be implemented using esterification, etherification, halogenation, oxidation, alkaline treatment and silylation [6,15,17,18]. The well-known direct modification method are presented in Figure 2.4. This method appears to be much simpler than chemical grafting since no chemical reaction is involved.

Monomer grafting is accomplished by attaching the chain grafts on the polymer backbone with covalent bonds. Various commonly used

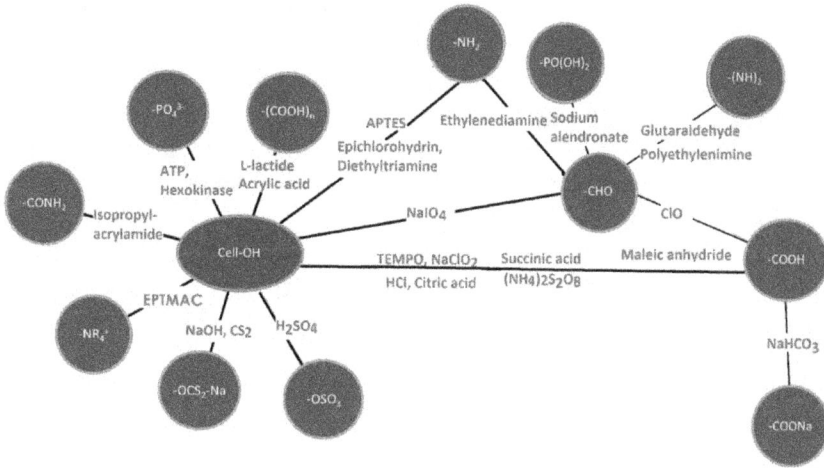

Figure 2.4 Direct chemical modification method.

techniques are photographing, high-energy radiation grafting and chemical initiation grafting [6,15,17,18]. Two main pathways can be used for grafting polymers on surfaces. These two strategies are schematically presented in Figure 2.5. The "grafting onto" procedure means that the cellulosic nanoparticles are mixed with an existing polymer and a coupling agent to attach the polymer to the nanoparticle surface. The "grafting from" approach involves the mixing of cellulosic nanoparticles with a monomer and an initiator to induce the polymerization of the monomer from the nanoparticle surface.

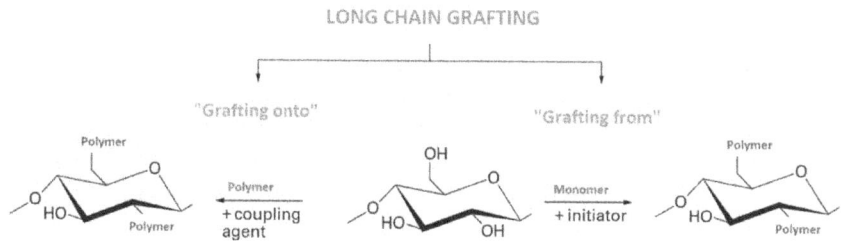

Figure 2.5 Representation of grafting approaches.

2.4.2 Composite Materials

Composites are multiphasic heterogeneous materials with improved properties as compared to matrix phase. Matrix phase is a continuous phase while reinforcement is are discontinuous in nature. Individual components in the resultant composites remain bonded together by physical or chemical means while retaining their physical identities.

Nanocellulose is an ideal matrix for a range of reinforcement nanomaterials due to its high specific surface area, porous structure and mechanical strength. Nanocellulose-based nanocomposites combine the advantages of both reinforcement nanomaterial and nanocellulose substrate, thereby, resulting in synergistic properties. There are three commonly used methods for incorporating the reinforcement in the nanocellulose substrate: 1) direct addition or formation of the reinforcement material in a nanocellulose dispersion; 2) formation of the reinforcement material in the bulk structure of nanocellulose-based materials and 3) direct coating of the reinforcement material onto the nanocellulose surface as a nano-sized layer. A range of different reinforcement nanomaterials such as metal and metal oxide nanoparticles (Fe, Au, Ag, Pd, Ni, TiO_2, CuO), mineral nanomaterials ($CaCO_3$, SiO_2, montmorillonite, bentonite), carbonaceous nanomaterials (carbon nanotube, graphene), etc., have been incorporated into or onto nanocellulose substrates [6,13,19].

The hydroxyl groups and porosity on the surface of nanocellulose results in an excellent mechanical support for catalysts, ensuring the dispersion of inorganic reinforcement nanoparticles and stability of catalysts, simultaneously.

2.4.3 Flitters and Membranes

Membrane separation technology is renowned method for water treatment due to its high efficiency, low energy consumption, less space requirement, good operability and absence of secondary pollution. The efficiency of nanocellulose-based membranes is mainly related to the high surface area nanocellulose fibers to selectively adsorb contaminants from water streams. The most reported processing techniques for development of nanocellulose membranes in the literature are: 1) impregnation of electrospun mats with nanocellulose; 2) vacuum filtration and coating and; 3) freeze-drying, with or without a matrix phase [20].

Nano-sized cellulosic materials are highly useful for the fabrication of polymer nanocomposites to improve the overall performance (mechanical and other functional properties, viz. partial degradation, barrier performance, etc.). Depending on the preparation methods, processing of polymer nanocomposite can be broadly divided into two categories: physical methods and chemical methods. Solution processing, melt mixing methods and electrospinning are typical examples of physical methods, whereas *in-situ* polymerization method can be considered as chemical method for the nanocomposite preparation [14].

2.4.4 Nanocellulose-based Nanostructured Photocatalyst

The assembly of photocatalysts such as TiO_2, Ag/TiO_2, Cu_2O, graphene oxide and ZnO onto nanocellulose surface promotes a new functionality of cellulose towards the development of bio-hybrid materials for various applications, especially in water treatment [21]. The interplay between the photocatalyst and cellulose surface chemistry is the main factor supporting the development of excellent association. The abundance of hydroxyl groups permits the assembly of photocatalyst on the cellulose surface. There are different methods concerning the preparation of photocatalyst nanomaterials within cellulose matrix such as sol-gel, hydrothermal, electrospinning, plasma vapor deposition, atomic layer deposition and hot press [21-26].

2.4.5 Nanocellulose-based Aerogels

Nanocellulose particles have the strong tendency of agglomerate in order to reduce their surface energy due to the surface hydroxyl groups of the cellulose fibrils. The cellulose fibrils promote their self-aggregation through hydrogen bonding. This property leads to a significant decrease in the adsorption capacity inherent to their high specific surface area. The use of nanocellulose in the form of a solid aerogel with desired shape is an alternative to alleviate this shortcoming [27]. Aerogels refer to a new class of nanoscale mesoporous materials with an open structure having a highly porous solid of ultra-low density and high surface area with nanometric pore sizes formed by replacement of liquid in a gel with gas. In addition to replacement of liquid with gas, the main challenge during the manufacturing process is to remove the solvent or disperse the liquid without collapse or shrinkage of the network structure under the effect of the strong capillary forces, especially when solvents have high surface energy, such as water or alcohol [28]. Supercritical drying is the most frequently used technique for the removal of solvent. Nevertheless, this method is not cost effective and not practical for high-volume production. An alternative procedure used in the production of nano-cellulose-based aerogels is freeze-drying. The idea of this method is to freeze the solvents first and subsequently sublimate without turning into a liquid state.

2.5 Applications of Nanocellulose-based Materials for Water Treatment

2.5.1 Removal of Heavy Metals

Heavy metals are naturally occurring elements with atomic weight and density at least 5 times greater than that of water. Due to toxic effects, non-biodegradability and the tendency to accumulate in the food chain, heavy metals are kept under environmental pollutant category [29,30]. As heavy metals are found throughout the earth's crust, environmental contamination can occur through metal corrosion, atmospheric deposition, soil erosion of metal ions, leaching of heavy metals, sediment re-suspension and metal evaporation from water resources to soil and ground water [6,30]. However, the maximum environmental contamination and human exposure result from anthropogenic activities such as mining and smelting operations, industrial production and use, domestic and agricultural applications of metals and metal-containing compounds, etc. [29,30]. In addition, many of the heavy metals, such as zinc, copper, chromium, iron and manganese, are essential in very small amounts for various body functions. However, serious health disadvantages can occur if these metals accumulate in the body at high concentrations. The heavy metals most commonly associated with poisoning are lead, mercury, arsenic and cadmium. Exposure to these metals has been associated with serious consequences such as developmental retardation, various cancers, kidney damage, auto-immunity and even death in extreme cases [31].

Only few studies had been reported on the heavy metal adsorption properties of nanocellulose before 2012. However, more studies have become available in the last few years. Chemically modified cellulose, cellulose based hybrid or composite materials, flocculants and membranes are observed to be worthy materials to remove heavy metals from water [2,6,11,15,20,32,33]. As maintained earlier, several nano-cellulose functionalization methods such as carboxylation, amination, succinylation and sulfonation have been employed to improve the adsorption efficiency of targeted metal ions. The major binding groups for metal ions are presented in Table 2.1.

Some examples of heavy metal removal with different nanocellulose-based materials are presented in Table 2.2. Arsenic, cadmium, chromium, lead and mercury rank among the priority metals that are of public health significance, thus, more attention has been paid for the removal of these elements.

Lead (Pb) is one of the most deleterious metal detected in waste streams from mining operations, tanneries, electronics, electroplating and petrochemical industries. Pb(II) removal from water by different nanocellulose-based adsorbents has been studied. For example, the use of sulfonated NC, nanocellulose/poly(2-(dimethylamino)ethyl methacrylate) interpenetrating polymer network hydrogels, tannin-nanocrystalline cellulose composite, rice straw nanocellulose fibers, carboxylated NC, nanofibrous microfiltration membranes, diethylenetriamine modified NC, carboxylated NC-filled magnetic chitosan hydrogel, magnetic nanocellulose, poly(methacrylic acid-co-maleic acid) grafted nanofibrillated cellulose, etc., has been reported [11,20,34-42]. Removal capacity of these materials varied form 1.2 mg/g to 485 mg/g.

Inorganic arsenic is a confirmed carcinogen and is one of the most significant chemical contaminants in drinking water globally. It is naturally present at high levels in the ground water of a number of countries, such as Argentina, Bangladesh, Chile, China, India, Mexico, and the United States of America. Arsenic is one of WHO's 10 chemicals of major public health concern. Conventionally applied techniques to remove arsenic include oxidation, coagulation-flocculation and membrane separation. Besides, the use of various nanoparticles for the remediation of arsenic contaminated water has been of interest recently. Particularly, iron-based adsorption is an emerging technique for the treatment of arsenic-contaminated water due to the high affinity between inorganic arsenic species and iron. Iron-based nanoparticles were also studied as component of NC composite materials for removal of arsenic from water [19,43]. Other nanocellulose-based absorption materials reported in literature for arsenic removal are $ZnO:CeO_2$:nanocellulose:polyaniline bio-nanocomposite and dopamine conjugated NC [44,45].

Cadmium (Cd) is a highly toxic and non-essential heavy metal in the environment. The main reason for cadmium pollution is rapid industrialization and other human activities. The industries/industrial operations generating cadmium waste include metallurgy, alloying, electroplating, photography, cadmium nickel batteries, phosphate fertilizers, pigmenting, textile printing, lead mine drainage, pesticides and dyes industries. Due to the non-biodegradable nature of Cd(II) ions, these are bioaccumulated and bioconcentrated in the living tissues. Toxicity of cadmium causes adverse health effects such as diarrhoea, pulmonary problems, renal damage, hypertension, anaemia, bone lesions, cancer, lung insufficiency and malfunctioning of liver, kidneys and reproductive organs [46]. According to the guidelines of World Health Organization, the allowable limit of Cd in drinking water is 0.003 mg/L.

The broad spectrum of different NC based adsorption materials has been studied for the removal of Cd(II) from water, with the achievement of a range of removal capacities. TEMPO-oxidation of carboxylic groups yield the maximum uptake of 9.7 mg/g [11]. Chemically modification of NC using succinic anhydride to form SCNCs (succinic anhydride cellulose nanocrystals) and sodium bicarbonate ($NaHCO_3$) enabled the maximum uptake of 344.8 mg/g for Cd(II). Cellulose nanofibers functionalized with oxolane-2,5-dione have been used for the adsorption of Pb(II) and Cd(II) from modeled waste water. The adsorption capacity of this nanomaterial was recorded to be 325.9 mg /g for Cd(II) [47]. Xanthated nanoscale banana cellulose (XNBC) was used for the adsorption of Cd(II) from aqueous solutions [48]. The sulfur groups in the adsorbent resulted in high uptake of Cd(II) ions (156.26 mg/g). The maximum removal capability of 262.27 mg/g for Cd(II) was reported by using synthesized P(MB-IA)-g-MNCC, a cellulose based adsorbent, via itaconic acid polymerization onto magnetic NCC [49].

Mercury (Hg) is one of the most harmful and ubiquitous pollutants in the global environment. Due to its toxicity, persistence in the environment and bioaccumulation, Hg has been considered as one of the most toxic metals which can affect the human health. It can be toxic in the form of organic as well as inorganic compounds. Long-term exposure to high concentrations of Hg can harm the human brain, heart, kidneys, lungs and even the immune system. Exposure to Hg in the womb or through seafood containing methyl mercury (MeHg) may harm the developing nervous system of unborn babies and young children [46]. Several methods have been used for the removal of Hg(II) ions such as chemical precipitation, electrochemical separation, ion exchange membrane filtration, solvent extraction, reverse osmosis and adsorption. Among these methods, adsorption is considered the most effective technique for the removal of Hg(II) as other methods are limited by excessive time requirements, high costs and production of highly toxic sludge [50]. Over the last few years, NC based materials have been developed for this purpose with promising results. For example, absorbent composed of cellulose with thiosemicarbazide Hg(II) reached the maximum capacity of 505.50 mg/g [51]. NC synthesized via acid hydrolysis, followed by lipase catalyzed esterification using 3-mercaptopropionic acid, was reported to have maximum adsorption capacity of 98.6 mg/g [52]. L-Cysteine functionalized bagasse cellulose nanofibers exhibited a maximum adsorption capacity of 116.822 mg/g [53]. NC aerogel containing both thiol and carboxyl groups exhibited a removal capacity of 718.5 mg/g [54].

Chromium is the seventh most abundant element on earth. It exists in several oxidation states, but the most stable and common forms are Cr(0), Cr(III) and Cr(VI) species. The toxicity of Cr in plants depends on its valence state: Cr(VI), as being highly mobile, is toxic, while Cr(III), as less mobile, is less toxic [55]. Cr (III) is oxidized to Cr (VI) because of the presence of excess oxygen in the environment. This form of chromium is extremely toxic and highly soluble in water, with mutagenic properties. Hence, Cr(VI) is categorized as a group 1 human carcinogen by the International Agency for the Research on Cancer [56]. For these reasons, the removal of chromium from water has received a significant attention. Its removal by NC based materials has also been studied widely. High adsorption capacity of 112.2 mg/g was achieved by using hydroxyapatite-NC composite material [57]. Dopamine conjugated NC demonstrated significant performance in water treatment for arsenic and Cr(VI) (25.95 mg/g), and almost equivalent capacity was also achieved with Fe_3O_4/NC composite material [44, 58]. Succinated and aminated nanocellulose exhibited capacity of 10.21 mg/g for Cr(III) as well as 11.84 mg/g for Cr(VI) [59].

2.5.2 Removal of Organic Pollutants

Organic pollution is the term used when large quantities of organic compounds are discharged in environment. Most of these compounds are toxic and include bio-refractory organics such as pesticides, fertilizers, hydrocarbons, phenols, plasticizers, biphenyls, detergents, oils, greases, pharmaceuticals, proteins and carbohydrates. In the following, the uptake of sparingly soluble toxic organic pollutants as well as dyes, oils and drugs by functionalized nanocellulose has been reviewed in detail.

Removal of Harmful or Toxic Organic Pollutants

Only limited number of studies have focused on the usefulness of nanocellulose-based materials as adsorbents for sparingly water soluble organic pollutants. The adsorption of organic molecules onto nanocellulose surface has been further promoted by grafting suitable organic moieties, which boost the interaction with organic molecules through van der Waals interactions. The grafting of linear aliphatic chains on nanocellulose surface was observed to strongly enhance the adsorption of sparingly water soluble organic compounds, including solvents, pesticides and herbicides [60]. The cellulose nanofibers in form of aerogels succeeded to adsorb 2-naphtol, nitrobenzene, xylene, bromobenzene and phenol.

Humic substances (HSs), like humic acid, are formed by decomposition of plant and animal biomass. Humic acid can cause various environmental and health problems including undesirable color and taste, binding to heavy metals and biocides which yield new toxic materials, reaction with chlorine to form potentially carcinogenic compounds, reduction of adsorption rate and equilibrium capacity of other compounds, etc. [61]. Due to a variety of functional groups in humic acid, different adsorption mechanisms (e.g., electrostatic, hydrogen bonding, hydrophobic forces, etc.) might be involved. Despite this, their removal is faced with some challenges. For example, activated carbon is not suitable for humic acid due to the high molecular weight and heavy metal oxide of zeolite and montmorillonite may be released during adsorption [62]. Considering this, nanocellulose-based adsorption materials are worthy alternative for removal of humic acid from water solutions. For example, amine-modified nanocellulose may be used for the removal of humic acid from waste water [62].

Removal of Dyes

Synthetic dyes are widely used in many industrial sectors such as tanneries, pharmaceuticals, paper, paints, plastics and cosmetics. Most of the organic dyes are toxic and non-biodegradable, with even teratogenetic, carcinogenic and mutagenic characteristics, which pose serious hazards for human and marine organisms. Hence, it is important to identify efficient dye removal techniques to overcome the ecological, biological and industrial challenges. Organic dyes usually have a complex aromatic structure and can indicate cationic, anionic or non-ionic properties. A commonly used treatment method is adsorption, due to its high efficiency and non-toxic nature. Moreover, adsorbents are readily available and show durable performance. Foer this purpose, nanocellulose-based adsorbents have been reported in the form of aerogels, membranes and nanocomposites.

Carboxylated nanocellulose has been extensively studied for the sorption of cationic dyes. For instance, carboxylated nanocellulose with a high COOH content, prepared using TEMPO-mediated oxidation, exhibited a maximum adsorption capacity of 769 mg/g for methylene blue, which was much superior as compared to the material with sulfate groups on the surfaces (118 mg/g) [63]. In another study, carboxylic acid functionalized nanocellulose, prepared by esterification of surface hydroxyl groups with maleic anhydride, displayed a high uptake capacity for several cationic dyes, e.g., crystal violet, methylene blue, malachite green and basic fuchsin [64]. Carboxylated nanocellulose prepared by using ammonium persulfate [65] and hydrochloric and citric acid [66] was also observed to adsorb methylene blue effectively. Nanocellulose-based crosslinked microgels with polyvinylamine (PVAm) exhibited a high affinity for both cationic and anionic dyes, with maximum adsorption uptakes for acid red GR, Congo red 4BS and reactive light yellow K-4G of 896 mg/g, 1469 mg/g and 1250 mg/g respectively [67]. Novel nanocomposite adsorbent membranes were developed by mixing chitosan matrix with nanocellulose, followed by freeze-drying and compression molding [68]. The membranes were able to remove almost 100% of Victoria blue, 90% of methyl violet and 80% of rhodamine 6G. More examples about dye removal by nanocellulose-based materials are presented in Table 2.2.

Removal of Contaminants of Emerging Concern

Pharmaceuticals, typically at levels ranging from nanograms to low micrograms per litre, have been reported in the water cycle, including surface waters, waste water, ground water and, to a lesser extent, drinking water [69]. Pharmaceuticals can contaminate water sources through sewage containing the excreta of patients using these chemicals, uncontrolled drug disposal (e.g. discarding drugs into toilets) and agricultural runoff comprising livestock manure. Pharmaceuticals have become chemicals of emerging concern to the public because of their potential to reach drinking water sources in the recent years.

Though a large number of research studies have reported the use of nanocellulose as drug carrier, studies on the adsorption of drugs onto nanocellulose are sparse. It is reported that sulfated NC can adsorb ionized drugs such as doxorubicin hydrochloride (DOX) and tetracycline hydrochloride TC in water [70,71]. Carboxylated NC also exhibited comparable binding capacities for cationic drugs [72]. Though only a few research studies have been are reported on this subject, however, it is clear that nanocellulose-based adsorbents can be efficient adsorbents of different drugs.

Removal of Oil

The probability of oil spill accidents has dramatically increased over the past several decades due to the growth of offshore oil production and transportation. The conventional methods currently employed for oil spill cleanup involve: a) dispersing the oil phase in water with the help of dispersing agents to facilitate natural degradation, b) burning the floating oil *in-situ* and c) extracting the oil phase from the water surface using a sorbent. Among these approaches, the removal of oil by sorbent materials is considered to be the most efficient, economical and ecologically friendly due to the effective removal of pollutant and absence of any secondary pollution [73]. Several characteristics are required for the oil adsorbent material, including a high oil absorption capacity and oil/water selectivity, fast oil sorption rate, high floatability (hence low density), low cost, environmentally friendliness and recyclability.

Aerogels are highly interconnected porous and lightweight solid materials formed by replacing the liquid in a gel by air. Their large specific surface area, high porosity and low density enable these aerogels to rapidly absorb a large amount of oil and to float on water, which makes them as an ideal oil sorbent candidate. Aerogels based on cellulose nanofibrils have been of great interest as absorbents for oil removal [27,73-75]. For example, the hydrophobic aerogels showed high absorption capacities for various oils, depending on liquid density, up to 47 times their original weight, but with low water uptake (<0.5 g/g aerogel) [75]. In another study, the volume-based and mass-based sorption capacities of hydrophobic (and oleophilic) nanocellulose aerogels reached 80-90% of total volume of the aerogel and 20-40 g/g [76]. Similarly, a series of highly porous (99.1-99.8%) and lightweight (4-14 mg/cm^3) nanocellulose aerogels was fabricated by freeze-drying aqueous carboxymethylated nanocellulose dispersions with varying solid contents. The hydrophobic aerogels filled almost instantly with the oil phase while selectively absorbing n-hexadecane from water, with a sorption capacity 45 times their own weight [77]. It was also found that the octyltrichlorosilane (OTCS) treatment had a minor negative effect on the absorption capacity of the aerogels toward heptane. In another study, natural pine needles were used to extract cellulose, which was then used to isolate aqueous nanocellulose suspensions by combining acid-pretreatment with high-intensity ultrasonic treatment. The material could absorb oil 52 times oil its own weight without any structural collapse [78].

2.5.3 Bacterial Removal

It is well known that microbial contamination in water is a global health hazard. The presence of microbial contaminants in ground and surface water sources is attributed to a number of factors, including the intrinsic physical and chemical characteristics of the catchment area and the magnitude and range of human activities and animal sources that release pathogens to the environment. Antibacterial activity of nanocellulose-based materials has been reported in various studies [79-81]. For water treatment purpose, nanocellulose composite fibrous membranes, consisting of an ultra-fine cellulose nanofibrous network infused into an electrospun polyacrylonitrile (PAN) nanofibrous scaffold on a melt-blown polyethylene terephthalate (PET) non-woven substrate, were prepared [82]. The developed materials exhibited excellent water filtration properties due to the controlled pore size in the barrier layer to sieve bacteria and positive surface charge to adsorb viruses (negatively charged). Nanocellulose-silver nanoparticles (AgNPs) embedded pebbles-based composite material was observed to have the high removal capacity of dyes, heavy metals and Escherichia coli from the simulated contaminated water [83]. NC-AgNPs in polyvinyl alcohol (PVA) exhibited removal capacity of 96.9% and 88.2% against gram-negative Escherichia coli and gram-positive Staphylococcus aureus, respectively [23].

2.5.4 Removal of Anionic Pollutants

A number of anionic species, such as nitrates, nitrites, chlorides, sulfides, fluorides and cyanides, may be considered as pollutants when present in water bodies above certain concentrations. Moreover, their bioaccumulation becomes an environmental concern when their concentrations in the environment begins to affect human health and ecosystems.

Ferric hydroxide-coated cellulose nanofibers were synthesized as composite adsorbents for the removal of phosphate from waste water [84]. The maximum sorption capacity for phosphate was estimated to be 142.86 mg/g. Cationic nanocellulose-based materials are efficient adsorbents of negatively-charged water contaminants, such as nitrates, phosphates, fluorides and sulfates. It was observed that the cationic nanocellulose displayed a higher selectivity toward multivalent ions (PO_4^{3-} and SO_4^{2-}) than monovalent ions (F^- and NO_3^-) [85]. The adsorption of PO_4^{3-}, NO_3^-, sodium sulfate (Na_2SO_4) and sodium lauryl sulfate (SLS) by calcium hydroxyapatite-nanocellulose composite was studied in the aqueous solution [86]. The maximum removal capacities were observed to be 75.98, 12.96 7.35 , 34.53 mg/g, respectively.

2.5.5 Flocculant for Waste Water Effluents

Coagulation and flocculation are the traditional ways to treat waste water. These processes are easy to handle and present relatively inexpensive ways to achieve solid-liquid separation. During the coagulation process, the colloids or particles in a suspension are destabilized, causing the formation of small aggregates. The resulting destabilized and aggregated particles or other colloidal and particular materials are gathered into larger aggregates during the flocculation process.

The usefulness of nanocellulose-based materials as a flocculent for the treatment of municipal or industrial waste water has been studied in the recent years. The anionic nanocelluloses for combined coagulation–flocculation treatment of municipal waste water show that chemically modified anionic dicarboxyl acid nanocelluloses are effective green alternatives for synthetic flocculants [32]. The anionic sulfonated nanocellulose was also tested in combined coagulation–flocculation treatment of municipal waste water with better performance than the aforementioned study [33]. Carboxylated nanocellulose was also used as efficient flocculant for kaolin suspensions with a turbidity removal of 99.5% at an initial kaolin dosage of 40 mg/L [66]. The flocculation efficiency was observed to be superior, as compared to commercial cationic polyacrylamide flocculants. The flocculation mechanism using anionic nanocellulose has also been illustrated in Figure 2.6.

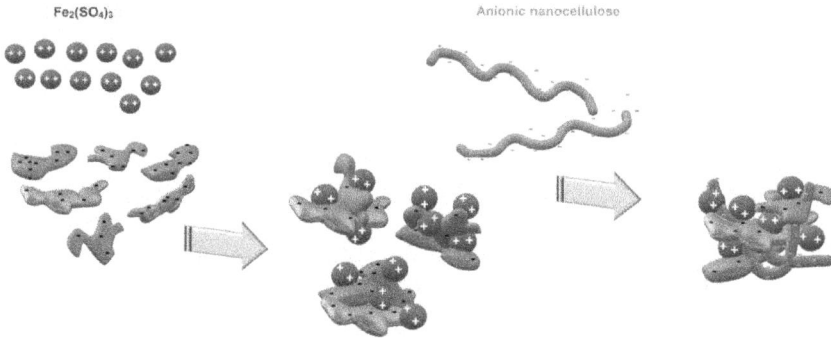

Figure 2.6 Illustration of the flocculation mechanism [11].

2.6 Adsorbent Regeneration

The reusability of adsorbents significantly affects the operational cost, thus, the regeneration process of adsorbents is a crucial parameter for the potential exploitation of these materials from an economic point of view. In case of nanocellulose-based adsorbents, special attention needs to be paid on this aspect as the manufacturing costs are higher in comparison with conventional adsorbents, due to the steps involving the production of nanocellulose and subsequent chemical modification. A simple, effective and cost-efficient method of regeneration will undoubtedly contribute towards promoting the potential use of nanocellulose-based adsorbents in real water treatment applications. An effective regeneration process desorbs the adsorbed compounds entirely without degradation of the properties of the adsorbent. Metal-recovery by desorption processes also allows the re-use of recovered metals, which is beneficial especially in the case of valuable or rare metals. Due to these reasons, regeneration of nanocellulose-based adsorption materials has received a significant research attention. As Table 2.3 illustrates, depending on the adsorbed species and modification method of nanocellulose, a multitude of methods have been proposed and tested for the adsorbent regeneration. In general, an acidic or alkaline treatment was used for heavy metal ion desorption [9,87]. Desorption of organic pollutants was conducted with an organic solvent [15,87]. In addition, the recyclability of nanocellulose-based materials has been tested by using sorption/desorption cycles, typically ranging between 3 and 5. Recyclable nanocellulose-based adsorbent materials exhibiting 80-90%

retention of adsorption capacity have been demonstrated with different material systems discussed in the earlier sections [38,40,58,59,76,86,88-90].

2.7 Current Challenges and Future Perspectives

Nanocellulose is a promising material as bio-based nanosized adsorbent for environmental remediation. However, some shortcomings need to be considered before these alternative adsorbents can be used as commercial adsorbents for water purification or waste water treatment process.

The cost effectiveness is the key challenge as nanocellulose materials are expensive to produce. However, recent investments by several companies in the paper-making industry (e.g. Exilva, Borregaard Sarpsborg, Norway; CELISH/TIARA, Diacel FineChem Ltd., Tokyo, Japan; UPM Kymmene Ltd. Helsinki, Finland; StoraEnzo Ltd., Stockholm, Sweden; Celluforce, Quebec, Canada; MoRE Research, Örnsköldsvik, Sweden; Forest Products Lab., Madison, WI, USA, etc.) for the cost-effective industrial production of nanocellulose is expected to provide the market with large amounts of nanocellulose. Expanding applications in the manufacturing of synthetic polymers and chemicals along with large-scale industrial production will in future lead to a reduction in the price of nanocellulose. According to the Global Industry Analysts, the market of nanocellulose-based materials is projected to exceed 1 billion U.S. dollars by 2020 [91]. In addition to the production of cost-efficient nanocellulose raw material, developing cost-efficient processing routes will be indispensable for the commercial implementation of nanocellulose-based adsorption materials for water treatment applications. Considering the environmental perspective, safe and cost-efficient disposal routes at the end of service life are essential for market realization of alternative products for water treatment.

The other limitation for commercial implementation of nanocellulose-based materials for water treatment is the lack of studies about the adsorption properties under dynamic conditions, more specifically for the treatment of industrial waste water. Even though numerous studies have been conducted in order to identify alternative adsorbents for water purification and waste water treatment process, most of these are limited to batch scale only. Thus, more research is required to demonstrate the utility of nanocellulose-based materials for real-life waste water treatment processes, as these processes are typically very complicated and various types of contaminants usually coexist. More investigations are also needed for the removal of multiple coexisting pollutants using real-life waste water as solution model. Study of the adsorption selectivity in the presence of multiple species is particularly important. In addition to the laboratory scale, studies need to be conducted in pilot plant scale. More attention should be paid to the development of environmental-friendly processes without using toxic chemical agents.

Even though the biodegradability of cellulose nanomaterials is an advantage in many applications considering the eco-friendliness perspective, however, it can lead to limitations and challenges in certain circumstances. Resistance to biological degradation is needed when nanocellulose materials interact with bacteria in water. The chemical modification and stabilization with different polymers are expected to reduce/inhibit the biological degradation of nanocellulose, however, the subject requires further research.

2.8 Conclusions

Various research studies have focused on the adsorption of hazardous pollutants using nanocellulose and its modified forms as adsorbents or membranes. As a renewable and biodegradable natural material, nanocellulose has high strength and flexibility, along with a high surface area and versatile surface chemistry. Nanocellulose is used as a raw material for the development of chemically modified adsorbents, composite materials, membranes and photocatalysts for water treatment. For nanocellulose-based adsorbents, the surface modification is a key step to promote the removal of a specific class of pollutants from water. A wide range of surface modification strategies has been observed to result in desired properties needed for the removal of different contaminants from waste water. The cationic or anionic surface groups of nanocellulose have been shown to remove heavy metal pollutants from aqueous solutions. The ionic and non-ionic surface groups have also been used to remove organic pollutants, e.g., dyes, oils and pesticides. Antibacterial activity of nanocellulose-based materials has also been investigated for water treatment purposes. Although nanocellulose-based materials have demonstrated the effectiveness towards the adsorption of numerous contaminants from water, their ultimate success as commercial products requires further development of cost-effective industrial production of nanocellulose.

Table 2.1 The major binding groups for metal ions

Functional group and Modification method of NC	Conta-minant	Q_{max} (mg/g)	Ref.
-COOH (carboxylation)	Cu(II)	20	[92]
	Pb(II)	65	
-COOH, -NH$_2$ (poly(vinyl alcohol) (PVA) hydrogel)	Pb(II)	211	[93]
None (CNF-OH obtained through mechanical treatment)	Cd^{2+}	11	[37]
	Ni$^+$	11	
	Pb$^+$	10	
-SO$_3^-$ (sulfuric acid hydrolysis)	Ag$^+$	34	[94]
-COOH (via NaIO$_4$–NaClO$_2$ oxidation)	Cu(II)	38	[95]
	Pb(II)	158	
-PO$_4^{2-}$ (phosphorylation of NC)	Ag+	136	[96]
	Cu^{2+}	117	
	Fe^{3+}	115	
-COOH (Succinic anhydride)	Pb^{2+}	458	[38]
	Cd^{2+}	335	
-COOH (sodium periodate/chlorite)	Cu^{2+}	185	[97]
-COOH (TEMPO)	Cu^{2+}	112	[88,98,99]
	Ni^{2+}	49	
	Cr(III)	58	
	Zn^{2+}	67	
	UO$_2^{2+}$	167	
	Cd(II)	10	
	Pb(II)	9	
	Ni(II)	9	
	Cr(II)	9	
NC-Fe$_3$O$_4$ -composite	Pb(II)	66	[58]
	Mn(II)	33	
	Cr(III)	25	
CNF-(PO(OH)$_2$)$_2$	VO$_3^-$	194	[100]
-NH$_2$ (ethylenediaminetetraacetic acid (EDTA))	Cd(II)	92	[101]
	Cu(II)	149.0	
	Pb(II)	333.0	

	Ni(II)	179	
-NH₂ (reaction with APTES)	Cu(II)	163	[17]
	Cd(II)	388	
-ROCS₂—Na (reaction with C2)	Cd(II)	154	[48]
-NH₂ (reaction with K₂S₂O₈ and ethylenediamine)	Cr(VI)	3	[59]
-NH₂ (reaction with epichlrorhydrin and diethylenetramine)	Pb²⁺	84	[39]
	Cu²⁺	63	
-NH₂ (grafting with PEG-NH₂)	Hg²⁺		[102]
-CONH₂ (grafting of isopropylacrylamide)	Hg²⁺		[103]
TEMPO -PEI	Cu²⁺	52	[104]
-CMC (crosslinked using butanetetracarboxylic acid)	Ag⁺	106	[105]
	Cu²⁺	75	
	Pb²⁺	112	
	Hg²⁺	131	

Table 2.2 Heavy metal removal with different nanocellulose-based materials

Functional group and Modification method of NC	Contaminant	Q_{max} (mg/g)	Ref.
Carboxylic groups (COO–)	Methylene blue (MB)	122.2	[106]
Amination (-NH₂)	Acid red dye	134.7	[68]
	Yellow K-4G dye	184.0	
	Congo red 4BS	199.5	
Sulfate ester groups (SO₃-) and carboxylic (COO-) groups	Methylene blue (MB)	769.0	[63]
Sulphate groups (SO₃⁻)	Methylene blue	79-85%	[107]
(-NH₂) (grafted with PVAm)	Congo red 4BS	869.1	[67]

	Acid red GR	1469.7	
	Reactive light yellow K-4G	1250.9	
CNC-NH$_2$ (oxidation with NaIO$_4$ followed by grafting with hyperbranched PEI)	Congo red	2100	[89]
	Basic yellow	1860	
CoCO$_3$/FeCO$_3$	Reactive Red 195 (RR195)	19	[108]
CNF-NR4+ (reaction with epoxypropyltrime-thylammonium chloride)	Congo Red	664.0	[109]
	Acid green 25	683.0	
Carboxylate-functionalized (COO-)	Crysta Violet (CV)	221.8	[64]

Table 2.3 Methods proposed for the regeneration of adsorbents

Type of contaminant	Type of modification	Regeneration method	Ref.
Cu^{2+}, Pb^{2+}	Aminated NC	EDTA	[92]
Cr^{6+}, Pb^{2+}	Polyacrylonitrile modified NC	EDTA	[110]
As^{5+}	Magnetic iron nanoparticles/NC	1.0 M NaOH	[19]
Ni^{2+}, Cu^{2+}, Cd^{2+}	Amino-propyl-triethoxy-silane modification	0.1 M NaOH	[17]
Au^{3+}	Sulfonated NC	0.5 M thiourea in 1 M HCl	[111]
Ni^{2+}, Cu^{2+}, Cd^{2+}	Fe$_3$O$_4$/NC	0.1 M NaOH	[112]
Mn^{2+}, Pb^{2+}, Cr^{3+}	Cellulose nanocrystals from rice straw	0.1 M HNO$_3$	[37]
PO$_4^{3-}$, NO$_3^-$	Hydroxyapatite/NC	0.1 M NaOH	[86]
Hg^{2+}	Thiolated NC	0.1M HCl	[52]
Methylene blue dye	NC/silver nanoparticles (AgNPs)	2 M HCl and 0.1 M NaOH	[83]
Oil	Silylated NC Sponges	Toluen	[74]

References

1. Schwarzenbach, R. P., Escher, B. I., Fenner, K., Hofstetter, T. B., Johnson, C. A., von Gunten, U., and Wehrli, B. (2006) The challenge of micropollutants in aquatic systems, *Science*, **313**, 1072-1077.
2. Gupta, V. K., Carrott, P. J. M., Carrott, M. M. L. R., and Suhas, (2009) Low-cost adsorbents: growing approach to wastewater treatment -

A review. *Critical Reviews in Environmental Science and Technology*, **39**, 783-842.

3. *Progress on Drinking Water, Sanitation and Hygiene: 2017 Update and SDG Baselines*, UNICEF (2017). Online: https://www.unicef.org/publications/index_96611.html [accessed 15th January 2019].

4. *Water: A Shared Responsibility; the United Nations World Water Development Report 2*, UNESDOC (2006). Online: https://unesdoc.unesco.org/ark:/48223/pf0000144409 [accessed 15th January 2019].

5. Shannon, M. A., Bohn, P. W., Elimelech, M., Georgiadis, J. G., Marinas, B. J., and Mayes, A. M. (2008) Science and technology for water purification in the coming decades, *Nature*, **452**, 301-310.

6. Hokkanen, S., Bhatnagar, A., and Sillanpää, M. (2016) A review on modification methods to cellulose-based adsorbents to improve adsorption capacity. *Water Research*, **91**, 156-173.

7. Weber, W. J., McGinley, P. M., and Katz, L. E. (1991) Sorption phenomena in subsurface systems: Concepts, models and effects on contaminant fate and transport. *Water Research*, **25**, 499-528.

8. Ruthven, D. M. (1984) *Principles of Adsorption and Adsorption Processes*, Wiley, USA.

9. Putro, J. N., Kurniawan, A., Ismadji, S., and Ju, Y. -H. (2017) Nanocellulose based biosorbents for wastewater treatment: Study of isotherm, kinetic, thermodynamic and reusability. *Environmental Nanotechnology, Monitoring & Management*, **8**, 134-149.

10. Lin, N., Huang, J., and Dufresne, A. (2012) Preparation, properties and applications of polysaccharide nanocrystals in advanced functional nanomaterials: a review. *Nanoscale*, **4**, 3274-3294.

11. Mahfoudhi, N., and Boufi, S. (2017) Nanocellulose as a novel nanostructured adsorbent for environmental remediation: A review. *Cellulose*, **24**, 1171-1197.

12. Dufresne, A. (2013) Nanocellulose: A new ageless bionanomaterial. *Materials Today*, **16**, 220-227.

13. Wei, H., Rodriguez, K., Renneckar, S., and Vikesland, P. J. (2014) Environmental science and engineering applications of nanocellulose-based nanocomposites. *Environmental Science: Nano*, **1**, 302-316.

14. Mondal, S. (2018) Review on nanocellulose polymer nanocomposites. *Polymer-Plastics Technology and Engineering*, **57**, 1377-1391.

15. O' Connell, D. W., Birkinshaw, C., and O'Dwyer, T. F. (2008) Heavy metal adsorbents prepared from the modification of cellulose: A review. *Bioresource Technology*, **99**, 6709-6724.

16. Nguyen, T., Roddick, F. A., and Fan, L. (2012) Biofouling of water treatment membranes: a review of the underlying causes, monitoring techniques and control measures. *Membranes*, **2**, 804-840.

17. Hokkanen, S., Repo, E., Suopajärvi, T., Liimatainen, H., Niinima, J.,

and Sillanpää, M. (2014) Adsorption of Ni(II), Cu(II) and Cd(II) from aqueous solutions by amino modified nanostructured microfibrillated cellulose. *Cellulose*, **21**, 1471-1487.

18. Hokkanen, S., Repo, E., and Sillanpää, M. (2013) Removal of heavy metals from aqueous solutions by succinic anhydride modified mercerized nanocellulose. *Chemical Engineering Journal*, **223**, 40-47.

19. Hokkanen, S., Repo, E., Lou, S., and Sillanpää, M. (2015) Removal of arsenic(V) by magnetic nanoparticle activated microfibrillated cellulose. *Chemical Engineering Journal*, **260**, 886-894.

20. Voisin, H., Bergström, L., Liu, P., and Mathew, A. P. (2017) Nanocellulose-based materials for water purification. *Nanomaterials*, **7**, 57.

21. Mohamed, M. A., Abd Mutalib, M., Mohd Hir, Z. A., Zain, M. F. M., Mohamad, A. B., Minggu, J. L., Awang, N. A., and Salleh, W. N. W. (2017) An overview on cellulose-based material in tailoring bio-hybrid nanostructured photocatalysts for water treatment and renewable energy applications. *International Journal of Biological Macromolecules*, **103**, 1232-1256.

22. Mohamed, M. A., Salleh, W. N. W., Jaafar, J., Ismail, A. F., Mutalib, M. A., Sani, N. A. A., Asri, S. E. A. M., and Ong, C. S. (2016) Physicochemical characteristic of regenerated cellulose/N-doped TiO2 nanocomposite membrane fabricated from recycled newspaper with photocatalytic activity under UV and visible light irradiation. *Chemical Engineering Journal*, **284**, 202-215.

23. Xu, X., Yang, Y. -Q., Xing, Y. -Y., Yang, J.-F., and Wang, S.-F. (2013) Properties of novel polyvinyl alcohol/cellulose nanocrystals/silver nanoparticles blend membranes. *Carbohydrate Polymers*, **98**, 1573-1577.

24. Wang, S., Luo, T., Zhu, J.,Zhang, X., and Su, S. (2016) A facile way to fabricate cellulose-Ag@AgCl composites with photocatalytic properties. *Cellulose*, **23**, 3737-3745.

25. Kemell, M., Pore, V., Ritala, M., Leskelä, M., and Lindén, M. (2005) Atomic layer deposition in nanometer-level replication of cellulosic substances and preparation of photocatalytic TiO2/cellulose composites. *Journal of the American Chemical Society*, **127**, 14178-14179.

26. Shinde, S. M., and Bhole, K. S. (2015) Review of Accuracy Improvement Techniques in High Speed 5 Axis Machining, *2015 International Conference on Nascent Technologies in the Engineering Field (ICNTE)*, pp. 1-5.

27. Tarrés, Q., Oliver-Ortega, H., Llop, M., Pèlach, M. À., Delgado-Aguilar, M., and Mutjé, P. (2016) Effective and simple methodology to produce nanocellulose-based aerogels for selective oil removal. *Cellulose*, **23**, 3077-3088.

28. Lavoine, N., and Bergstrom, L. (2017) Nanocellulose-based foams and aerogels: processing, properties, and applications. *Journal of*

Materials Chemistry A, **5**, 16105-16117.

29. Fergusson, J. E. (1990) *The Heavy Elements: Chemistry, Environmental Impact And Health Effects*, Pergamon Press, UK.
30. Tchounwou, P. B., Yedjou, C. G., Patlolla, A. K., and Sutton, D. J. (2012) Heavy metals toxicity and the environment. *EXS*, **101**, 133-164.
31. Järup, L. (2003) Hazards of heavy metal contamination. *British Medical Bulletin*, **68**, 167-182.
32. Suopajärvi, T., Liimatainen, H., Hormi, O., and Niinimäki, J. (2013) Coagulation–flocculation treatment of municipal wastewater based on anionized nanocelluloses. *Chemical Engineering Journal*, **231**, 59-67.
33. Suopajärvi, T., Koivuranta, E., Liimatainen, H., and Niinimäki, J. (2014) Flocculation of municipal wastewaters with anionic nanocelluloses: Influence of nanocellulose characteristics on floc morphology and strength. *Journal of Environmental Chemical Engineering*, **2**, 2005-2012.
34. Suopajärvi, T., Liimatainen, H., Karjalainen, M., Upola, H., and Niinimäki, J. (2015) Lead adsorption with sulfonated wheat pulp nanocelluloses. *Journal of Water Process Engineering*, **5**, 136-142.
35. Li, J., Xu, Z., Wu, W., Jing, Y., Dai, H., and Fang, G. (2018) Nanocellulose/Poly(2-(dimethylamino)ethyl methacrylate) Interpenetrating polymer network hydrogels for removal of Pb(II) and Cu(II) ions. *Colloids and Surfaces A: Physicochemical and Engineering Aspects*, **538**, 474-480.
36. Xu, Q., Wang, Y., Jin, L., Wang, Y., and Qin, M. (2017) Adsorption of Cu (II), Pb (II) and Cr (VI) from aqueous solutions using black wattle tannin-immobilized nanocellulose. *Journal of Hazardous Materials*, **339**, 91-99.
37. Kardam, A., Raj, K. R., Srivastava, S., and Srivastava, M. M. (2014) Nanocellulose fibers for biosorption of cadmium, nickel, and lead ions from aqueous solution. *Clean Technologies and Environmental Policy*, **16**, 385-393.
38. Yu, X., Tong, S., Ge, M., Wu, L., Zuo, J., Cao, C., and Song, W. (2013) Adsorption of heavy metal ions from aqueous solution by carboxylated cellulose nanocrystals. *Journal of Environmental Sciences*, **25**, 933-943.
39. Shen, W., Chen, S., Shi, S., Li, X., Zhang, X., Hu, W., and Wang, H. (2009) Adsorption of Cu(II) and Pb(II) onto diethylenetriamine-bacterial cellulose. *Carbohydrate Polymers*, **75**, 110-114.
40. Zhou, Y., Fu, S., Zhang, L., Zhan, H., and Levit, M. V. (2014) Use of carboxylated cellulose nanofibrils-filled magnetic chitosan hydrogel beads as adsorbents for Pb(II). *Carbohydrate Polymers*, **101**, 75-82.
41. Wang, R., Guan, S., Sato, A., Wang, X., Wang, Z., Yang, R., Hsiao, B. S., and Chu, B. (2013) Nanofibrous microfiltration membranes capable of removing bacteria, viruses and heavy metal ions. *Journal of Mem-*

brane Science, **446**, 376-382.

42. Maatar, W., and Boufi, S. (2015) Poly(methacylic acid-co-maleic acid) grafted nanofibrillated cellulose as a reusable novel heavy metal ions adsorbent. *Carbohydrate Polymers*, **126**, 199-207.

43. Taleb, K., Markovski, J., Veličković, Z., Rusmirović, J., Rančić, M., Pavlović, V., and Marinković, A. (2016) Arsenic removal by magnetite-loaded amino modified nano/microcellulose adsorbents: Effect of functionalization and media size. *Arabian Journal of Chemistry*, doi:10.1016/j.arabjc.2016.08.006.

44. Dwivedi, A. D., Sanandiya, N. D., Singh, J. P., Husnain, S. M., Chae, K. H., Hwang, D. S., and Chang, Y. -S. (2017) Tuning and characterizing nanocellulose interface for enhanced removal of dual-sorbate (AsV and CrVI) from water matrices. *ACS Sustainable Chemistry & Engineering*, **5**, 518-528.

45. Nath, B. K., Chaliha, C., Kalita, E., and Kalita, M. C. (2016) Synthesis and characterization of ZnO:CeO2:nanocellulose:PANI bionano-composite. A bimodal agent for arsenic adsorption and antibacterial action. *Carbohydrate Polymers*, **148**, 397-405.

46. *Guidelines for Drinking-water Quality: Fourth Edition Incorporating the First Addendum*, World Health Organization, Switzerland (2017).

47. Stephen, M., Catherine, N., Brenda, M., Andrew, K., Leslie, P., and Corrine, G. (2011) Oxolane-2,5-dione modified electrospun cellulose nanofibers for heavy metals adsorption. *Journal of Hazardous Materials*, **192**, 922-927.

48. Pillai, S. S., Deepa, B., Abraham, E., Girija, N., Geetha, P., Jacob, L., and Koshy, M. (2013) Biosorption of Cd(II) from aqueous solution using xanthated nano banana cellulose: Equilibrium and kinetic studies. *Ecotoxicology and Environmental Safety*, **98**, 352-360.

49. Anirudhan, T. S., and Shainy, F. (2015) Adsorption behaviour of 2-mercaptobenzamide modified itaconic acid-grafted-magnetite nanocellulose composite for cadmium(II) from aqueous solutions. *Journal of Industrial and Engineering Chemistry*, **32**, 157-166.

50. Anirudhan, T. S., and Shainy, F. (2015) Effective removal of mercury(II) ions from chlor-alkali industrial wastewater using 2-mercaptobenzamide modified itaconic acid-grafted-magnetite nanocellulose composite. *Journal of Colloid and Interface Science*, **456**, 22-31.

51. Zhou, Y., Hu, X., Zhang, M., Zhuo, X., and Niu, J. (2013) Preparation and characterization of modified cellulose for adsorption of Cd(II), Hg(II), and acid fuchsin from aqueous solutions. *Industrial & Engineering Chemistry Research*, **52**, 876-884.

52. Ram, B., and Chauhan, G. S. (2018) New spherical nanocellulose and thiol-based adsorbent for rapid and selective removal of mercuric ions. *Chemical Engineering Journal*, **331**, 587-596.

53. Bansal, M., Ram, B., Chauhan, G. S., and Kaushik, A. (2018) l-Cysteine functionalized bagasse cellulose nanofibers for mercury(II) ions adsorption. *International Journal of Biological Macromolecules*, **112**, 728-736.

54. Geng, B., Wang, H., Wu, S., Ru, J., Tong, C., Chen, Y., Liu, H., Wu, S., and Liu, X. (2017) Surface-tailored nanocellulose aerogels with thiol-functional moieties for highly efficient and selective removal of Hg(II) ions from water. *ACS Sustainable Chemistry & Engineering*, **5**, 11715-11726.

55. Kimbrough, D. E., Cohen, Y., Winer, A. M., Creelman, L., and Mabuni, C. (1999) A critical assessment of chromium in the environment. *Critical Reviews in Environmental Science and Technology*, **29**, 1-46.

56. Jaishankar, M., Tseten, T., Anbalagan, N., Mathew, B. B., and Beeregowda, K. N. (2014) Toxicity, mechanism and health effects of some heavy metals. *Interdisciplinary Toxicology*, **7**, 60-72.

57. Hokkanen, S., Bhatnagar, A., Repo, E., Lou, S., and Sillanpää, M. (2016) Calcium hydroxyapatite microfibrillated cellulose composite as a potential adsorbent for the removal of Cr(VI) from aqueous solution. *Chemical Engineering Journal*, **283**, 445-452.

58. Zhu, H., Jia, S., Wan, T., Jia, Y., Yang, H., Li, J., Yan, L., and Zhong, C. (2011) Biosynthesis of spherical Fe3O4/bacterial cellulose nanocomposites as adsorbents for heavy metal ions. *Carbohydrate Polymers*, **86**, 1558-1564.

59. Singh, K., Arora, J. K., Sinha, T. J. M., and Srivastava, S. (2014) Functionalization of nanocrystalline cellulose for decontamination of Cr(III) and Cr(VI) from aqueous system: computational modeling approach. *Clean Technologies and Environmental Policy*, **16**, 1179-1191.

60. Maatar, W., Alila, S., and Boufi, S. (2013) Cellulose based organogel as an adsorbent for dissolved organic compounds. *Industrial Crops and Products*, **49**, 33-42.

61. Giasuddin, A. B., Kanel, S. R., and Choi, H. (2007) Adsorption of humic acid onto nanoscale zerovalent iron and its effect on arsenic removal. *Environmental Science & Technology*, **41**, 2022-2027.

62. Jebali, A., Behzadi, A., Rezapor, I., Jasemizad, T., Hekmatimoghaddam, S. H., Halvani, G. H., and Sedighi, N. (2015) Adsorption of humic acid by amine-modified nanocellulose: An experimental and simulation study. *International Journal of Environmental Science and Technology*, **12**, 45-52.

63. Batmaz, R., Mohammed, N., Zaman, M., Minhas, G., Berry, R. M., and Tam, K. C. (2014) Cellulose nanocrystals as promising adsorbents for the removal of cationic dyes. *Cellulose*, **21**, 1655-1665.

64. Qiao, H., Zhou, Y., Yu, F., Wang, E., Min, Y., Huang, Q., Pang, L., and Ma, T. (2015) Effective removal of cationic dyes using carboxylate-functionalized cellulose nanocrystals. *Chemosphere*, **141**, 297-303.

65. He, X., Male, K. B., Nesterenko, P. N., Brabazon, D., Paull, B., and Luong, J. H. T. (2013) Adsorption and desorption of methylene blue on porous carbon monoliths and nanocrystalline cellulose. *ACS Applied Materials & Interfaces*, **5**, 8796-8804.

66. Yu, H. -Y., Zhang, D. -Z., Lu, F. -F., and Yao, J. (2016) New approach for single-step extraction of carboxylated cellulose nanocrystals for their use as adsorbents and flocculants. *ACS Sustainable Chemistry & Engineering*, **4**, 2632-2643.

67. Jin, L., Sun, Q., Xu, Q., and Xu, Y. (2015) Adsorptive removal of anionic dyes from aqueous solutions using microgel based on nanocellulose and polyvinylamine. *Bioresource Technology*, **197**, 348-355.

68. Karim, Z., Mathew, A. P., Grahn, M., Mouzon, J., and Oksman, K. (2014) Nanoporous membranes with cellulose nanocrystals as functional entity in chitosan: Removal of dyes from water. *Carbohydrate Polymers*, **112**, 668-676.

69. *Pharmaceuticals in Drinking-water*, World Health Organization (2011). Online: http://www.who.int/water_sanitation_health/publications/2011/pharmaceuticals_20110601.pdf [accessed 11th January 2019].

70. Jackson, J. K., Letchford, K., Wasserman, B. Z., Ye, L., Hamad, W. Y., and Burt, H. M. (2011) The use of nanocrystalline cellulose for the binding and controlled release of drugs, *International Journal of Nanomedicine*, **6, 321**-330.

71. Rathod, M., Haldar, S., and Basha, S. (2015) Nanocrystalline cellulose for removal of tetracycline hydrochloride from water via biosorption: Equilibrium, kinetic and thermodynamic studies. *Ecological Engineering*, **84**, 240-249.

72. Akhlaghi, S. P., Tiong, D., Berry, R. M., and Tam, K. C. (2014)Comparative release studies of two cationic model drugs from different cellulose nanocrystal derivatives. *European Journal of Pharmaceutics and Biopharmaceutics*, **88**, 207-215.

73. Liu, H., Geng, B., Chen, Y., and Wang, H. (2017) Review on the aerogel-type oil sorbents derived from nanocellulose. *ACS Sustainable Chemistry & Engineering*, **5**, 49-66.

74. Zhang, Z., Sèbe, G., Rentsch, D., Zimmermann, T., and Tingaut, P. (2014) Ultralight weight and flexible silylated nanocellulose sponges for the selective removal of oil from water. *Chemistry of Materials*, **26**, 2659-2668.

75. Mulyadi, A., Zhang, Z., and Deng, Y. (2016) Fluorine-free oil absorbents made from cellulose nanofibril aerogels. *ACS Applied Materials & Interfaces*, **8**, 2732-2740.

76. Korhonen, J. T., Kettunen, M., Ras, R. H. A., and Ikkala, O. (2011) Hydrophobic Nanocellulose aerogels as floating, sustainable, reusable, and recyclable oil absorbents. *ACS Applied Materials & Interfaces*, **3**, 1813-1816.

77. Cervin, N. T., Aulin, C., Larsson, P. T., andWågberg, L. (2012) Ultra porous nanocellulose aerogels as separation medium for mixtures of oil/water liquids. *Cellulose*, **19**, 401-410.

78. S. Xiao, R. Gao, Y. Lu, J. Li, and Q. Sun, (2015) Fabrication and characterization of nanofibrillated cellulose and its aerogels from natural pine needles. *Carbohydrate Polymers*, **119**, 202-209.

79. Missoum, K., Sadocco, P., Causio, J., Belgacem, M. N., and Bras, J. (2014) Antibacterial activity and biodegradability assessment of chemically grafted nanofibrillated cellulose. *Materials Science and Engineering: C*, **45**, 477-483.

80. Bideau, B., Bras, J., Saini, S., Daneault, C., and Loranger, E. (2016) Mechanical and antibacterial properties of a nanocellulose-polypyrrole multilayer composite. *Materials Science and Engineering: C*, **69**, 977-984.

81. Jorfi, M., and Foster, E. J. (2015) Recent advances in nanocellulose for biomedical applications. *Journal of Applied Polymer Science*, **132**, doi:10.1002/app.41719.

82. Sato, A., Wang, R., Ma, H., Hsiao, B. S., and Chu, B. (2011) Novel nanofibrous scaffolds for water filtration with bacteria and virus removal capability. *Journal of Electron Microscopy*, **60**, 201-209.

83. Suman, Kardam, A., Gera, M., and Jain, V. K. (2015) A novel reusable nanocomposite for complete removal of dyes, heavy metals and microbial load from water based on nanocellulose and silver nano-embedded pebbles. *Environmental Technology*, **36**, 706-714.

84. Cui, G., Liu, M., Chen, Y., Zhang, W., and Zhao, J. (2016) Synthesis of a ferric hydroxide-coated cellulose nanofiber hybrid for effective removal of phosphate from wastewater. *Carbohydrate Polymers*, **154**, 40-47.

85. Sehaqui, H., Mautner, A., Perez de Larraya, U., Pfenninger, N., Tingaut, P., and Zimmermann, T. (2016) Cationic cellulose nanofibers from waste pulp residues and their nitrate, fluoride, sulphate and phosphate adsorption properties. *Carbohydrate Polymers*, **135**, 334-340.

86. Hokkanen, S., Repo, E., Westholm, L. J., Lou, S., Sainio, T., and Sillanpää, M. (2014) Adsorption of Ni^{2+}, Cd^{2+}, PO_4^{3-} and NO^{3-} from aqueous solutions by nanostructured microfibrillated cellulose modified with carbonated hydroxyapatite. *Chemical Engineering Journal*, **252**, 64-74.

87. Kwon, T. -N., and Jeon, C. (2012) Desorption and regeneration characteristics for previously adsorbed indium ions tophosphorylated sawdust. *Environmental Engineering Research*, **17**, 65-67.

88. Srivastava, S., Kardam, A., and Raj, K. R. (2012) Nanotech reinforcement onto cellulosic fibers: green remediation of toxic metals. *International Journal of Green Nanotechnology*, **4**, 46-53.

89. Zhu, W., Liu, L., Liao, Q., Chen, X., Qian, Z., Shen, J., Liang, J., and Yao,

J. (2016) Functionalization of cellulose with hyperbranched poly-ethylenimine for selective dye adsorption and separation. *Cellulose*, **23**, 3785-3797.

90. Gadd, G. M. (2009) Biosorption: critical review of scientific rationale, environmental importance and significance for pollution treatment. *Journal of Chemical Technology & Biotechnology*, **84**, 13-28.

91. *Global Nanocellulose Market*, Zion Market Research (2019). Online: https://www.zionmarketresearch.com/report/nanocellulose-market [accessed 19th January 2019].

92. Chen, S., Zou, Y., Yan, Z., Shen, W., Shi, S., Zhang, X., and Wang, H. (2009) Carboxymethylated-bacterial cellulose for copper and lead ion removal. *Journal of Hazardous Materials*, **161**, 1355-1359.

93. Hui, B., Zhang, Y., and Ye, L. (2015) Structure of PVA/gelatin hydrogel beads and adsorption mechanism for advanced Pb(II) removal. *Journal of Industrial and Engineering Chemistry*, **21**, 868-876.

94. Liu, P., Sehaqui, H., Tingaut, P., Wichser, A., Oksman, K., and Mathew, A. P. (2014) Cellulose and chitin nanomaterials for capturing silver ions (Ag+) from water via surface adsorption. *Cellulose*, **21**, 449-461.

95. Yao, C., Wang, F., Cai, Z., and Wang, X. (2016) Aldehyde-functionalized porous nanocellulose for effective removal of heavy metal ions from aqueous solutions. *RSC Advances*, **6**, 92648-92654.

96. Liu, P., Borrell, P. F., Božič, M., Kokol, V., Oksman, K., and Mathew, A. P. (2015) Nanocelluloses and their phosphorylated derivatives for selective adsorption of Ag^+, Cu^{2+} and Fe^{3+} from industrial effluents. *Journal of Hazardous Materials*, **294**, 177-185.

97. Sheikhi, A., Safari, S., Yang, H., and van de Ven, T. G. M. (2015) Copper removal using electrosterically stabilized nanocrystalline cellulose. *ACS Applied Materials & Interfaces*, **7**, 11301-11308.

98. Sehaqui, H., de Larraya, U. P., Liu, P., Pfenninger, N., Mathew A. P., Zimmermann, T., and Tingaut P. (2014) Enhancing adsorption of heavy metal ions onto biobased nanofibers from waste pulp residues for application in wastewater treatment. *Cellulose*, **21**, 2831-2844.

99. Ma, H., Hsiao, B. S., and Chu, B. (2012) Ultrafine cellulose nanofibers as efficient adsorbents for removal of UO_2^{2+} in water. *ACS Macro Letters*, **1**, 213-216.

100. Sirviö, J. A., Hasa, T., Leiviskä, T., Liimatainen, H., and Hormi, O. (2016) Bisphosphonate nanocellulose in the removal of vanadium(V) from water. *Cellulose*, **23**, 689-697.

101. Júnior, O. K., Gurgel, L. V. A., de Freitas, R. P., and Gil, L. F. (2009) Adsorption of Cu(II), Cd(II), and Pb(II) from aqueous single metal solutions by mercerized cellulose and mercerized sugarcane bagasse chemically modified with EDTA dianhydride (EDTAD). *Carbo-*

hydrate Polymers, **77**, 643-650.

102. Araki, J., Wada, M., and Kuga, S. (2001) Steric stabilization of a cellulose microcrystal suspension by poly(ethylene glycol) grafting. *Langmuir,* **17**, 21-27.

103. Geay, M., Marchetti, V., Clément, A., Loubinoux, B., and Gérardin, P. (2000) Decontamination of synthetic solutions containing heavy metals using chemically modified sawdusts bearing polyacrylic acid chains. *Journal of Wood Science,* **46**, 331-333.

104. Zhang, N., Zang, G.-L., Shi, C., Yu, H.-Q., and Sheng, G.-P. (2016) A novel adsorbent TEMPO-mediated oxidized cellulose nanofibrils modified with PEI: Preparation, characterization, and application for Cu(II) removal. *Journal of Hazardous Materials,* **316**, 11-18.

105. Chen, B., Zheng, Q., Zhu, J., Li, J., Cai, Z., Chen, L., and Gong, S. (2016) Mechanically strong fully biobased anisotropic cellulose aerogels. *RSC Advances,* **6**, 96518-96526.

106. Chan, C. H., Chia, C. H., Zakaria, S., Sajab, M. S., Chin, S. X. (2015) Cellulose nanofibrils: a rapid adsorbent for the removal of methylene blue. *RSC Advances,* **5**(24), 18204-18212.

107. Mohamed, M. A., Salleh, W. N. W., Jaafar, J., Ismail, A. F., Abd Mutalib, M., Mohamad, A. B., Zain, M. F. M., Awang, N. A., Mohd Hir, Z. A. (2017) Physicochemical characterization of cellulose nanocrystal and nanoporous self-assembled CNC membrane derived from Ceiba pentandra. *Carbohydrate Polymers,* **157**, 1892-1902.

108. Nassar, M. Y., and Khatab, M. (2016) Cobalt ferrite nanoparticles via a template-free hydrothermal route as an efficient nano-adsorbent for potential textile dye removal. *RSC Advances,* **6**(83), 79688-79705.

109. Pei, A., Butchosa, N., Berglund, L. A., and Zhou, Q. (2013) Surface quaternized cellulose nanofibrils with high water absorbency and adsorption capacity for anionic dyes. *Soft Matter,* **9**(6), 2047-2055.

110. Yang, R., Aubrecht, K. B., Ma, H., Wang, R., Grubbs, R. B., Hsiao, B. S., and Chu, B. (2014) Thiol-modified cellulose nanofibrous composite membranes for chromium (VI) and lead (II) adsorption. *Polymer,* **55**, 1167-1176.

111. Dwivedi, A. D., Dubey, S. P., Hokkanen, S., Fallah, R. N., and Sillanpää, M. (2014) Recovery of gold from aqueous solutions by taurine modified cellulose: An adsorptive–reduction pathway. *Chemical Engineering Journal,* **255**, 97-106.

112. Nata, I. F., Sureshkumar, M., and Lee, C.-K. (2011) One-pot preparation of amine-rich magnetite/bacterial cellulose nanocomposite and its application for arsenate removal. *RSC Advances,* **1**, 625-631.

3

Characteristics and Applications of Bacterial Nanocellulose

Alicia N. Califano,[a],* María Laura Balquinta,[a] Patricia Cerrutti,[b] Silvina C. Andrés[a] and Gabriel Lorenzo[a,c]

[a]*Centro de Investigación y Desarrollo en Criotecnología de Alimentos (CIDCA), CONICET, CICPBA, Facultad de Ciencias Exactas, UNLP. 47 y 116, La Plata (1900), Argentina*
[b]*Instituto de Tecnología en Polímeros y Nanotecnología (ITPN-UBA-CONICET), Facultad de Ingeniería, UBA, Las Heras 2214, Buenos Aires, Argentina and Departamento de Ingeniería Química, Facultad de Ingeniería, UBA, Argentina*
[c]*Departamento de Ingeniería Química, Facultad de Ingeniería, UNLP, Argentina*

Corresponding author: anc@quimica.unlp.edu.ar

3.1 Introduction

There is an increasing demand for products made from renewable and sustainable non-petroleum based resources. Cellulose, the most abundant polymer on earth, is renewable, biodegradable as well as non-toxic. In nature, cellulose is a ubiquitous structural polymer that confers its mechanical properties to higher plant cells.

Cellulose is a linear polymer of β-(1→4)-D-glucopyranose units in $^4C_1$ conformation. Therefore, the repeating unit of cellulose is known to comprise two anhydroglucose rings joined via a β-(1→4) glycosidic linkage (called cellobiose). The two end-groups of this polymer are not chemically equivalent, since one bears the "normal" C–OH group (non-reducing end), whereas the other has a C1–OH moiety in equilibrium with the corresponding aldehyde function (reducing end). It is a very common component of all vegetables as well as some algae and a few molds. The plants contain a combination of hemicellulose and lignin, as a ternary complex present in any cell wall [1]. In addition, there are some strains of the prokaryotic non-photosynthetic organisms, which have the ability to synthesize

Nanocellulose, edited by Vikas Mittal

high-quality cellulose organized as twisting ribbons of microfibril bundles [2]. The number of glucose units or the degree of polymerization (DP) in the polymer chain is influenced by the origin and treatment of the raw material.

Cellulose is an insoluble material consisting between 2000-14000 residues. It forms crystals (cellulose I_α) where intra-molecular (O3-H→O5' and O6→H-O2') and intra-strand (O6-H→O3') hydrogen bonds hold the network flat allowing the more hydrophobic ribbon faces to stack. Each residue is oriented 180° to the next with the chain synthesized two residues at a time. Although individual strands of cellulose are intrinsically no less hydrophilic, or no more hydrophobic, than other soluble polysaccharides (such as amylose), the tendency to form crystals utilizing extensive intra- and intermolecular hydrogen bonding makes it completely insoluble in normal aqueous solutions (although it is soluble in more exotic solvents such as N-methylmorpholine-N-oxide (NMNO) or LiCl/N,N'-dimethylacetamide). Water molecules are considered to be responsible for catalyzing the formation of the natural cellulose crystals by aligning the chains through hydrogen-bonded bridging.

Part of the cellulose structure between the crystalline sections is amorphous. The overall structure is of aggregated particles with extensive pores capable of holding relatively large amounts of water by capillarity. The natural crystal is made up from metastable cellulose I with parallel cellulose strands. This material can be divided into two coexisting phases: cellulose I_α and cellulose I_β, with differing displacements of the chains relative to one another. On recrystallization (e.g. from base or CS_2), it gives thermodynamically more stable cellulose II structure with an antiparallel arrangement of the strands and intra-sheet hydrogen-bonding to some extent.

Due to the hierarchical structure and semi-crystalline nature of cellulose, nanoparticles can be extracted from the natural polymer using a top-down mechanically or chemically induced deconstructing strategy or a bottom-up production of cellulose nanofibrils from glucose by bacteria. Cellulosic materials with one dimension in the nanometer range, such as whiskers, microfibrillated cellulose (MFC), nanofibrillated cellulose and cellulose nanofibrils or microfibrils [3,4] are generally referred to as nanocelluloses. These materials combine important properties of cellulose such as broad chemical-modification capacity, hydrophilicity and the formation of versatile semi-crystalline fiber morphologies. Nanocelluloses may be classified in three main subcategories: MFC (delamination of wood pulp by mechanical pressure before and/or after chemical or enzymatic treatment, 5-60 nm diameter, several micrometers length), nanocrystalline cellulose (NCC, acid hydrolysis of cellulose from many sources, 5-70 nm diameter, 100-250 nm length) and bacterial nanocellulose (BNC, bacterial synthesis, 20-100 nm diameter) [3]. While the diameters of NCC and BNC are similar, the main difference between the two types of nanocellulose is their purity and crystal structure. BNC is essentially pure cellulose and NCC is usually a composite itself consisting of both cellulose and hemicellulose [5,6].

3.2 Bacterial Nanocellulose

The discovery of BNC dates back to 1886 when Brown [7] described a acetic ferment as a jelly-like translucent mass on the surface of the culture fluid, commonly known as the "vinegar plant", composed of pure cellulose. Due to the ability to form cellulose, Brown [7] called this microorganism *Bacterium xylinum*, which was later renamed as *Acetobacter xylinum*. Nowadays, it is known as *Komagataeibacter xylinus*, following another designation as *Gluconacetobacter xylinus* and is classified to the genus *Komagataeibacter* [8,9].

These bacteria are wide-spread in nature where the fermentation of sugars and plant carbohydrates takes place. In contrast to MFC and NCC materials isolated from cellulose sources, BNC is formed as a polymer and nanomaterial by biotechnological assembly processes from low-molecular-weight carbon sources, such as D-glucose, using a bottom up method [10]. The bacteria are cultivated in common aqueous nutrient media, and the BNC is excreted as exopolysaccharide at the interface with air. The synthesis of BNC occurs between the outer and plasma membranes of the cell by cellulose synthesizing complex starting with UDP glucose. Cellulose synthase catalyzes the addition of UDP glucose to the end of the growing cellulose chain, which exits the cell as nanofibrils that are bundled in the form of ribbon-shaped microfibrils of *ca.* 80 nm x 4 nm size and subsequently form entangled structure in a tri-dimensional network with other cellulose fibrils. The resulting form-stable BNC hydrogel is composed of a nanofiber network enclosing up to 99% water. Figure 3.1 shows a partial view of bacterial nanocellulose sheet.

3.2.1 BNC Synthesis

Culture is processed in static or agitated conditions at temperatures

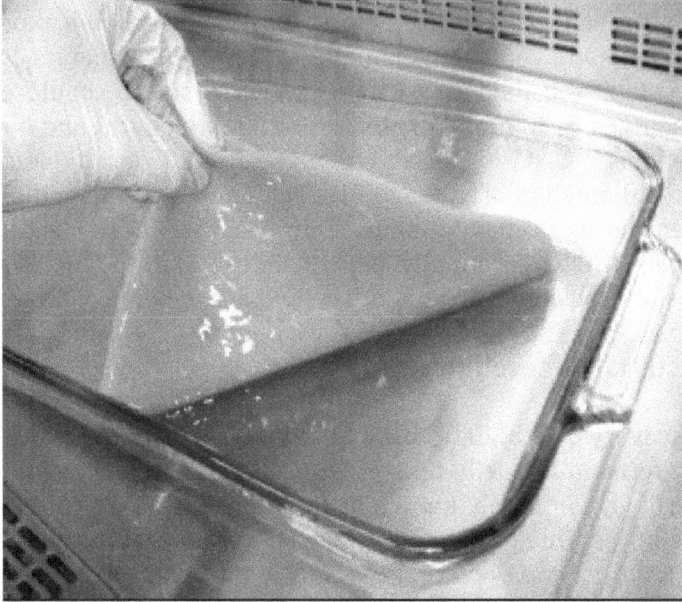

Figure 3.1 Partial view of bacterial nanocellulose sheet.

around 28-30 °C. BNC is secreted extracellularly by specific bacteria, mainly *Komagataeibacter* strains, which are Gram negative, aerobic and rodlike microorganisms (Figure 3.2). Owing to the aerobic nature of bacteria, in static fermentations, cellulose pellicle is formed only in the vicinity of the oxygen-rich air-liquid surface. The reason for bacterial generation of cellulose is unclear, however, it has been suggested that it is an approach that bacteria use to maintain their position close to the surface of culture solution [11] as well as to generate a protective coating that guards bacteria from ultraviolet radiation [12], prevents the entrance of contaminants and heavy-metal ions, whereas nutrients diffuse easily along the pellicle [13].

Thus, BNC seems to provide bacteria mechanical, chemical and biological protection, maintaining the cells trapped in the polymer liquid-air interface, a benefit for strictly aerobic microorganisms. The polymer matrix anchors the bacterial cells to the surface to allow for sufficient oxygen exposure and may also concentrate the nutrients by adsorption, favoring cell nutrition [14,15].

The effectiveness of microbial cellulose production depends mainly on the strain, composition of the culture medium (the C source is very significant), fermentation temperature, oxygen supply and implementation of static or agitated systems. The cost of media is an important factor for cost-effective BNC production, therefore, its composition and volume as well as the fermenter surface area should be considered [16].

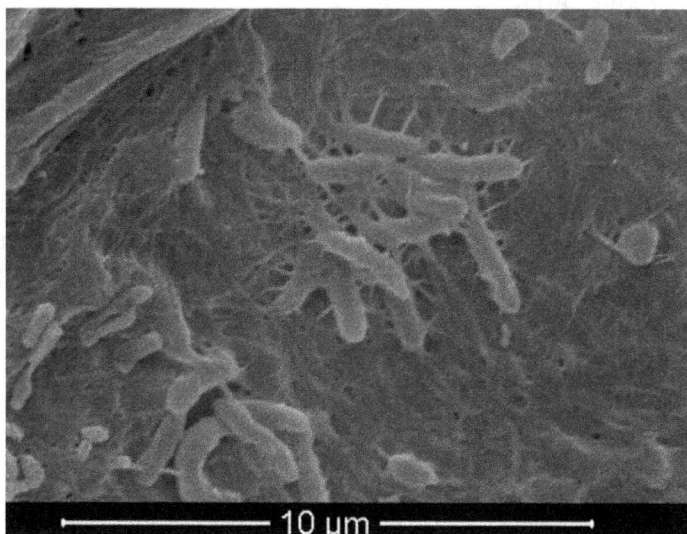

Figure 3.2 *Komagataeibacter* bacteria forming cellulose nanofibers and ribbons.

Hestrin-Schramm's medium (%, w/v: glucose 2.0, peptone 0.5, yeast extract 0.5, disodium phosphate 0.27 and citric acid 0.115) is generally used for the production of *G. xylinus* cellulose at laboratory scale [17]. However, it has been shown that *G. xylinus* can efficiently produce BNC from a wide range of other C sources such as fructose, sucrose, mannitol, maltose, xylose and glycerol [18,19]. In industrial processes, agro-industrial byproducts like cane molasses from manufacturing and refining of sugarcane, glycerol from biodiesel, vegetable oils, methanol, hydrocarbons, etc., are normally employed because of their lower costs [20-23]. Residual bacteria and components of the culture medium can be removed by heating in 0.05-0.10 M aqueous sodium hydroxide under reflux for 10-120 min, depending on the thickness of the cellulose body. Under these conditions, no detectable damage to the polymer occurs. Pure BNC is obtained with a yield higher than 40% (in relation to the bacterial strain), a value particularly elevated for a biotechnological route: it, in fact, does not contain impurities or functional groups other than the hydroxylic ones [24]. The efficiency of the process is strongly dependent on several factors such as the type of bacterial strain, surface structure of the substrate, component of the culture medium and temperature. Moreover, a continuous supply of oxygen and carbon source (D-glucose) is also required.

Flat products of different geometry are formed during static cultivation in liquid culture medium or in thin layer cultivation on solid phases like agar, silicone, rubber and different porous membranes. The size and thickness of the BNC fleeces and foils are influenced by the type of strain, volume of culture medium and cultivation time [25]. Hollow bodies of different shapes are formed by using a matrix in the static culture. Instead, BNC spheres or irregular shapes [26] can be obtained under agitated cultivation conditions (shaking, stirring, etc.).

3.2.2 Properties

Structure

BNC is proved to be very pure cellulose with a high weight average molecular weight, high crystallinity and good mechanical stability. The produced fibers are inherently nano-sized ribbon shaped cellulose fibrils, with the largest lateral dimension ranging from 25 to 86 nm and lengths of up to several micrometers [27], as can be seen in Figure 3.3.

Figure 3.3 Images of bacterial nanocellulose produced by *Gluconacetobacter xylinus* obtained by electronic scanning microscopy at different magnifications.

BNC has the same molecular formula as plant cellulose composed of β-(1$\rightarrow$4) -glucan chains (DP between 4000 and 10000 anhydroglucose units) that assemble to form crystalline and non-crystalline structures. The resulting form-stable BNC hydrogel is composed of an ultrafine nanofiber network structure (fiber diameter: 20-100 nm) enclosing up to 99% water [22,28-30]. This fine structure makes BNC different from other microbial polysaccharides, producing high water holding capacity, superior tensile strength, high purity and flexibility [13]. The high water holding capacity arises from the extensive surface area of nano- and microfibrils combined with its highly hydrophilic nature.

Fink *et al.* [31] elaborated a model of the BNC structure in which anhydrous nanofibrils in the range of 7 nm x 13 nm appear hydrated as a whole and are aggregated to flat microfibrils with a width of 70-150 nm (Figure 3.4). This indicates that water is outside of the crystalline cellulose nano-units. A shell of non-crystalline cellulose chains passes around neighboring microfibrils to produce a microfibril band (ribbon) with a width of about 0.5 µm. The incorporated water plays an important role as spacer element and stabilizing agent with respect to the network and pore structure. The assembly of stiffened cellulose chains in the crystals are held together by weak van der Waals forces overlying the adjacent sheets [32].

It is well known that BNC is a composite of the two crystalline polymorph phases of cellulose I, cellulose I_α (triclinic structure) and I_β (monoclinic structure, $I_\alpha/I_\beta = 65/35$) [33,34]. The difference between the allomorphs lies in the H-bonding systems and in the conformation of neighboring cellulose chains [25].

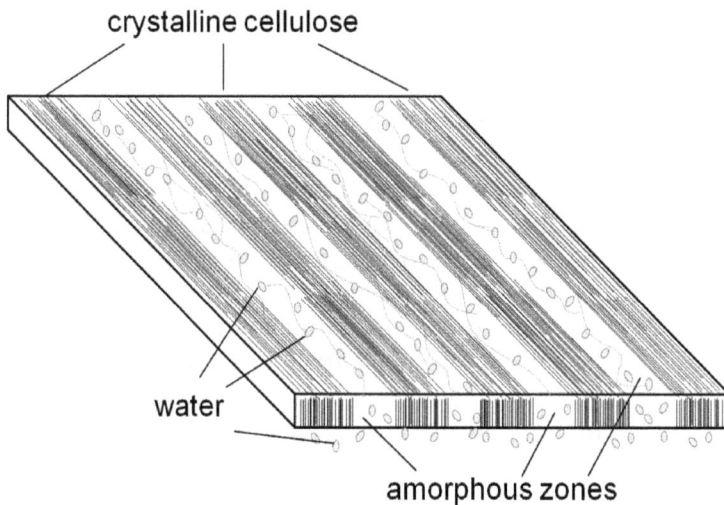

Figure 3.4 Model of hydrated bacterial nanocellulose fibrils.

Physicochemical Properties

BNC is wettable in never-dried state, and it can hold water tens to hundreds of times its weight. However, depending on the microorganism cultivation as culture medium, carbon source and other process conditions, structural changes could be observed, like aggregation of cellulose microfibrils modifying the crystallinity index as well as mechanical and electrical properties [15].

Several literature studies have reported thermal properties of bacterial cellulose using thermogravimetric analysis (TGA). Mostly, TGA curves of BNC show a first degradation stage due to moisture evaporation, which occurs between 25 °C and 130 °C [26,35]. The second weight loss is assigned to cellulose decomposition with onset and maximum temperatures between 300 °C and 305 °C [34]. Cerruti *et al.* [23] reported that the onset temperature of decomposition changed with pellicles age, and this could be attributed to the increased number of effective hydrogen bonds between BNC nanofibers with time. Extending the TG analysis up to 800 °C evidences a third smaller weight loss, between 450 and 600 °C. This high-temperature weight loss has been attributed to polymer chain degradation and the six-member cyclic structure, pyran, distinguishing it from the previous decomposition step assigned to the removal of molecular fragments such as hydroxyl and hydroxymethyl groups [36].

Figure 3.5 corresponds to a thermogram obtained by differential scanning calorimetry (DSC; heating rate 10 °C/min) of bacterial nanocellulose obtained from corn steep liquor + glycerol [37]. The heat flow curve shows an endothermic peak around 200 °C which can be related to the decomposition of cellulose to levoglucosan (1,6-anhydro-d-glucopyranose). Moreover, two exothermic peaks are also detected. The peak at 250 °C does not correspond to a mass loss and probably is due to the degradation of the intermolecular hydrogen bonds between BNC nanofibers. Kilzer and Broido [38] related this transition to the formation of anhydrocellulose. The other exothermic peak at 335 °C is attributed to the partial pyrolysis with fragmentation of carbonyl and carboxylic bonds from anhydrous glucose units producing carbon and/or carbon monoxide.

Another important property of cellulose and its derivatives is the degree of crystallinity. Wide-angle X-ray powder diffractometry (XRD) is a rapid analytical technique primarily used for phase identification of a crystalline material. The structure of bacterial cellulose obtained using different culture media was analyzed by XRD [22,35]. Diffraction patterns of BC obtained from glucose, commercial glycerol, glycerol remaining from biodiesel production, grape

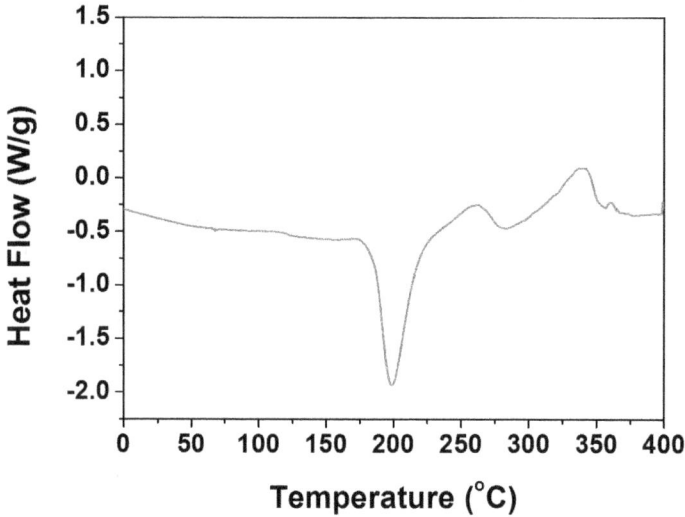

Figure 3.5 Differential scanning calorimetry thermogram of BNC produced from corn steep liquor and glycerol.

bagasse and cane molasses exhibited similar characteristics. Three peaks shown at 2θ = 14.62° (1-10, I_α), 16.29° (110) and 22.48° (200, I_β) confirmed that only cellulose I was present in all BC samples [39,40]. The position of these peaks is not in full agreement with theoretical values for the crystal planes (1-10), (110) and (200) at $2\theta \approx$ 14.88°, 16.68° and 22.9°, respectively. This peaks shift can be attributed to crystallite size variations which result in different long-range compressive forces on the crystals and unit cells.

Figure 3.6 shows a diffraction pattern of BNC obtained from corn steep liquor + glycerol and subsequently freeze dried to produce a nanocellulose powder by Balquinta *et al.* [37]. As also reported by other authors, characteristic peaks corresponding to (1-10), (110) and (200) planes are observed. Due to the intrinsic nature of polymerization and crystallization promoted by *Acetobacter xylinum*, the crystalline material formed is of type I and not of type II. Additionally, less intensity shoulder at 20.5° is also observed. This is in agreement with the (102) reflection expected for crystallites of cellulose I_β with random orientation [41,42].

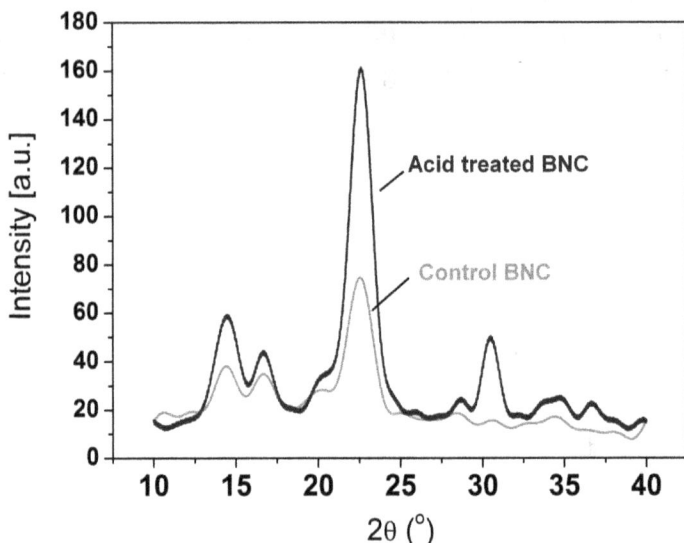

Figure 3.6 XRD pattern of BNC and BNC pretreated with hydrochloric acid.

The crystallinity index (CI) could be estimated using the height of the (200) peak (I_{200}) and the minimum intensity between the (200) and (110) peaks, which accounts for the amorphous part of cellulose (I_{am}):

$$CI(\%) = 100 \left(\frac{I_{200} - I_{am}}{I_{200}} \right)$$

Crystallinity determinations evidence that results obtained for different carbon sources are very similar; varying between 74-79% for BNC obtained from wine production residues, commercial glycerol or glucose, while the use of cane molasses apparently led to lower crystallinity, i.e. 67% [22].

Differences in crystallinity index are also detected when the obtained BNC was hydrolyzed with 2.5 M HCl at 70 °C for 2 h (Figure 3.6). Prior to freeze drying, the acid was removed with an initial partial neutralization with 0.1 M NaOH, followed by successive centrifugation and redispersion of the solid materials in ultrapure water. A significant difference was detected between the control BNC (69%) and the acid treated BNC (89%). Moreover, the treated BNC exhibited two additional peaks around 27.5° and 32°. These could be attributed to the 111 and 200 planes of the NaCl crystal that are present in the sample after neutralization.

Crystallite size perpendicular to each of the three main planes (L) and interplanar distance (*d*-spacing) for each crystallographic plane have also been reported [23,35]. The crystallite size is an indicative of the material stiffness and varies with the cellulose content [43]. The *d*-spacing values are important because these can provide information on the type of cellulose (I_α and I_β), since crystal structures are different (triclinic and monoclinic). The average crystallite size (L) could be estimated from the Scherrer equation:

$$L(nm) = \frac{K\lambda}{FWHM.cos(\theta)}$$

where, L is the crystallite size perpendicular to the plane, λ is the X-ray wavelength (Cu Kα = 0.1542 nm), *FWHM* is the full-width at half-maximum in radians and θ is the Bragg angle in radians.

Coelho de Carvalho Benini *et al.* [35] observed that the values of *d*-spacing obtained for nanoparticles from *Imperata brasiliensis* grass were similar to those obtained for microcrystalline cellulose (*Cladophora* sp.) treated with sulfuric acid 60% (w/w) [43] and for a structural model of cellulose I [44]. On the other hand, crystallite sizes were more sensitive to the hydrolysis conditions, particularly the (110) plane which contains more OH groups than the 1-10 plane, ranging from 5.28 nm to 8.39 nm. Cerruti *et al.* [23] reported that the estimated L value was around 5 nm for all BNC samples, regardless of fermentation time. Samples derived from sugarcane bagasse resulted with similar values of d and L, either with or without an acidic treatment, i.e. d $\approx$ 0.39 nm and L $\approx$ 5.8 nm.

Fourier transform infrared spectroscopy (FTIR) has also been used to characterize BNC [45-47]. A characteristic spectra can be seen in Figure 3.7, where a broad band in the region 3342 cm^{-1}, observed for all samples, indicates the O-H free stretching vibration of the CH_2-OH structure of cellulose [48] and OH groups which correspond to intra and intermolecular hydrogen bonds present in cellulose and absorbed water [49,50].

The peaks around 2900 cm^{-1} are due to C-H and CH_2 stretching vibration in both cellulose II and amorphous cellulose. The peaks located around 1636 cm^{-1} are attributed to bending vibrations of the OH groups of cellulose [51]. The bands observed at 1425 cm^{-1} are related to the C-C stretching and/or CH_2 symmetric bending in aromatic groups of cellulose due to crystalline structure [52-55]. Additionally, a peak at 1316 cm^{-1} was observed in both samples, and it could be attributed to the symmetric wagging of CH$_2$ (Oh *et al.*, 2005). The band at 1162 cm^{-1} is assigned to the antisymmetric bridge COC stretching mode. This peak appears in the infrared spectra of a number of celluloses and their derivatives, such as cellulose I, cellulose II, cellulose acetate, cellulose nitrate, 6-trityl cellulose, amylose, chitin, among others [56-58]. The band at 1106 cm^{-1} can be associated with ether C-O-C functionalities [59]. The absorbance band at 1056 cm^{-1} and 1032 cm^{-1} are associated to C-O stretching vibration of pyranose ring skeletal [51,59]. The band near 663 cm^{-1} is present in both samples and can be assigned to one of the C-OH out-of plane bending modes.

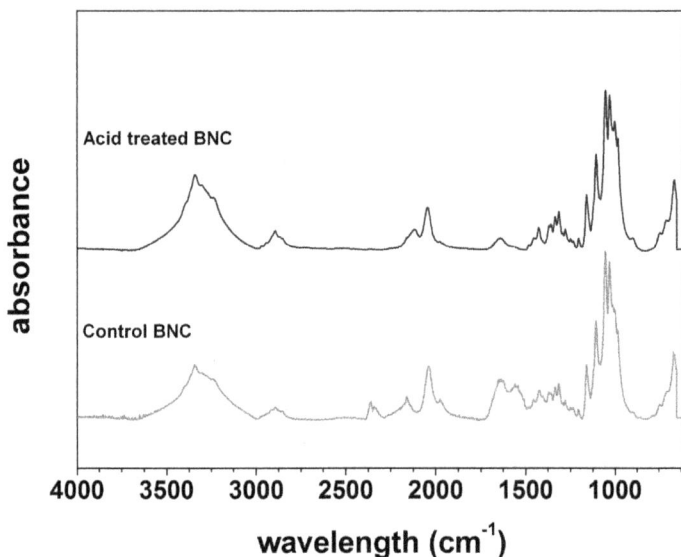

Figure 3.7 FTIR-ATR spectra of control and acid treated BNC.

Control BNC presented a peak at 1560 cm^{-1} that was not detected in the acid treated samples. Peaks in the range of 1500-1660 cm^{-1} are associated to proteins [48]. The presence of nitrogenated compounds in BNC pellicles can be attributed to nutrient medium residues remaining in the samples, as has been reported before [60]. The absence of this band in treated BNC could be related to the washing process to neutralize and remove the remaining HCl. The most significant difference in the FTIR spectra of the control BNC and the one pretreated with hydrochloric acid was observed 1636 cm^{-1} (Figure 3.7). Several literature studies have attributed this peak to adsorbed water [56,57,61]. It is noteworthy that this band tends to be weaker with increase in the crystallinity of the sample.

Besides, the –N$_3$ asymmetric stretching band appears around 2100 cm^{-1} [62-64], which corresponds to the azide added to the samples in order to prevent the appearance of molds during manipulation.

3.3 Applications

Bacterial nanocellulose is a biopolymer with a 3-D network structure [63]. Klemm *et al.* [3,25] provided extensive reviews on the biosynthesis of BNC and recent advances in various research areas such as advanced fiber composites, medical and tissue engineering. In 2013, Charreau *et al.* [65] found 815 patents available on uses of BNC, with a high extent of these dedicated to medical uses. The applications include (a) membranes for filtration as well as other separation processes such as dialysis and ion exchange [66], (b) medicinal pads for topical application made of a pellicle of microbially produced cellulose impregnated with a physiologically acceptable liquid [67], (c) molded materials of high dynamic strength formed from bacteria-produced cellulose comprising ribbon-like microfibrils [68], (d) oral plasters for treatment of stomatitis or gingivitis [69], (e) hydraulic fracturing fluids for recovery of hydrocarbons [70], (f) cartilage-like biomaterial designed for reconstructive surgery comprising purified and modeled BC [66], (g) hollow BNC useful in novel artificial blood vessels and in other medical material [71], (h) preparation of shaped articles of microbial cellulose useful in microsurgery by culturing microorganism between shape-defining walls [72], (i) coagulant for water treatment [73], among others.

3.3.1 BNC as Food Additive

BNC is a type of dietary fiber [74], classified as "generally recognized as safe" (GRAS), accepted as such by the USA Food and Drug Administration in 1992 [75]. BNC is a non-toxic hydrogel [76], and its rheological properties have led to its use in the food industry as a bacterial cellulose gel, called "nata de coco", which is eaten as a dessert delicacy since the 18th century. As a filler, BNC has the highest water holding capacity among commercial cellulose products, and it is considered a heat-stable stabilizer with low viscosity and a low calorie fat substitute. It can also be used as a food packing material because of its edibility and biodegradability [75].

It is used in food products, such as low-calorie desserts, salads and high-fiber foods. It is highly regarded for its high content of dietary fiber as well as low fat and cholesterol contents [76]. The 3D fibrillar network of BNC was also reported as stabilizer for cocoa particles in chocolate drinks [76]. BNC prevents the precipitation of the particles by producing a mesh that retains the particles in suspension. Moreover, after heat sterilization of food products, the viscosity of BNC-based food remains unchanged, showing a great thermal stability. Lin and Lin [77], for example, have developed a typical emulsified meat product (Chinese-style meatballs) containing BNC. BNC is a dietary fiber that offers a range of health benefits and can assist in reducing the risk of chronic diseases, such as diabetes, obesity, cardiovascular diseases and diverticulitis [78]. Important features of this application are a reduction in food retention in the intestine and an increase in bile acid secretion as well as a lowering of cholesterol and triacylglycerol levels in animal experiments. BNC can also serve as a stabilizer when added to ice cream. It has been reported that BNC is the best ingredient for ice cream to resist meltdown and heat shock [25]. Purwadaria *et al.* [79] mentioned the use of BNC and *Monascus* extract in vegetarian "meat". The complex is stable in color and morphology, and it tastes just like meat. It can also be used as a potential fat replacement in emulsified meat products. Marchetti *et al.* [10] described a novel application of BNC for the stabilization of low-lipid hypo-sodium meat emulsions formulated with pre-emulsified high-oleic sunflower oil. They concluded that using oil pre-emulsified with BNC showed significant potential to stabilize meat systems and as a fat mimetic to replace beef tallow in low-fat low-sodium meat sausages. Its use increased process yield, water-binding properties, hardness, cohesiveness and chewiness. The addition of BNC was not detrimental to shelf-life since quality parameters remained stable during the 45-days cold vacuum storage.

3.3.2 Further Applications of BNC

The use of additives like enzymes, emulsifiers, oxidants and hydrocolloids is a common practice to improve breadmaking performance and to preserve freshness and food properties. Bread can be enriched with dietary fiber, including wheat bran, gums (such as guar gum and modified cellulloses) and β-glucans. However, the addition of fiber to dough is a subject of controversy in the literature. They increase the total dietary fiber intake of the consumer and decrease the caloric density of baked breads [80]. However, from a technological and sensory point of view, it may cause decrease of loaf volume, increase in crumb firmness and dark crumb appearance [81]. Clearly, fiber in dough interacts with the gluten matrix. The extent of interaction as well as effect on the quality varies depending on the chemical structure of the added hydrocolloids [82].

Recently, Corral *et al.* [28] incorporated BNC into a typical formulation of French bread baked loaves containing dry yeast 1g/100 g flour, sugar (0.9 g/100 g flour), NaCl (2.6 g/100 g flour) and water (54.1 g/100 g flour); this formulation was considered as a control. Besides, a second formulation with 0.14 g of BNC (dry-basis)/100 g flour was prepared. BNC hydrogel was first dispersed in water and processed in a blender before adding the dispersion (0.25 g dry BNC/100 g water) to the rest of the components. Ingredients were mixed, rested, kneaded, divided into balls, molded into silicone muffin cups, proofed and finally baked. After the breads were baked, specific volume, yield, color, crumb characterization (by image analysis and confocal laser scanning microscopy (CLSM)) and textural properties (texture profile analysis) were measured to establish the effect of added BNC. Product yield was high (> 86.5 g/100 g dough) regardless of formulation. However, the addition of BNC increased specific volume of loaf and crumb moisture content. Larger pores appeared in BNC breads, thus, resulting in higher specific volume. Equivalent volume mean of the pores (D[4,3]) was 2.45 ± 0.08 mm for BNC samples, decreasing to 2.16 ± 0.02 mm for control bread crumb. Larger pores in the BNC breads produced a less dense and firm crumb, slightly less resilient and cohesive than the control (Figure 7.8).

Rheological principles and theory can be used as an aid in process control and design. These also act as a tool for the simulation and prediction of the material's response to the complex flows and deformation conditions often found in practical processing situations. Particularly, the dynamic oscillatory curves give a fingerprint of the state of the microstructure. When this type of measurement was applied on bread doughs, oscillatory stress sweeps presented a similar shear thinning behavior for both unleavened dough formulations, with G' >> G" and a small increase of the two moduli with frequency which corresponded to the characteristics of a weak gel. However, the strength of the crosslinked network was larger for the formulations containing BNC, implying the existence of stronger entanglements among hydrocolloids molecules in the composite network. Corral *et al.* [28] also calculated the plateau modulus ($G_N{}^0$) and the steady-state zero shear rate viscosity (η_0) from the discrete relaxation spectra. $G_N{}^0$ is a viscoelastic parameter defined for polymers as the extrapolation of the entanglement contribution to the viscoelastic functions at high frequencies and η_0 corresponds to the limit viscosity at very low values of shear rate. For the dough containing BNC, the obtained parameters were $G_N{}^0 = 15.3 \times 10^5$ Pa and $\eta_0 = 13.7 \times 10^5$ Pa.s, while for the control formulation, much smaller values were computed (7.41×10^5 Pa and 1.48×10^5 Pa.s, respectively), showing that BNC definitively reinforced the system.

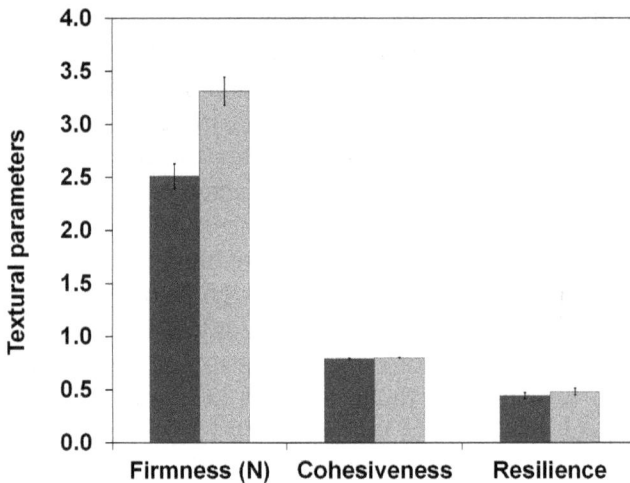

Figure 3.8 Textural parameters of French bread crumb: ■ BNC formulation (first column); ▨ control (second column).

The changes in dough microstructure that occur as a result of BNC inclusion were also reflected when observing the samples of dough and crumb with CLSM. Control dough exhibited a highly filamentous and oriented gluten network with a large number of entanglements (Figure 3.9a). The addition of BNC led to a less crosslinked matrix, with thicker and better aligned gluten filaments, with a larger separation between filaments. Control dough seemed more compact than the BNC formulation, that is, BNC samples exhibited a more open matrix (Figure 3.9b). Correa *et al.* [83] have reported a similar effect on the addition of HPMC to wheat bread doughs.

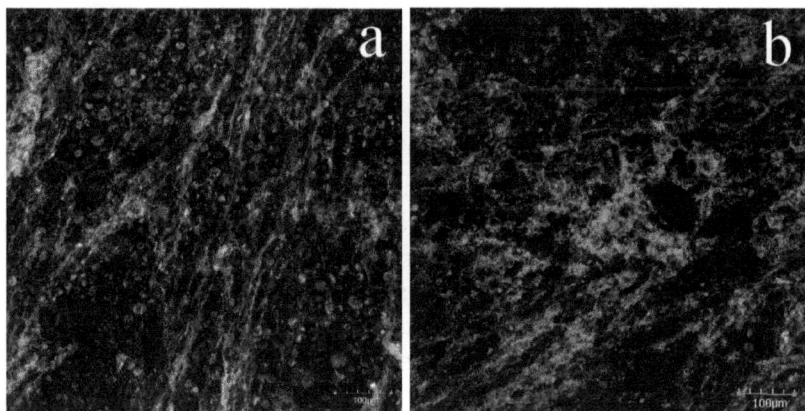

Figure 3.9 CLSM of bread dough: a) control without BNC and b) with bacterial nanocellulose. Black bar indicates 100 µm.

Figure 3.10 shows CLSM micrographs of both types of crumbs. Starch occupied the surface of the pore and the gluten filaments extended from side to side of the alveolus. Control crumb showed very thin strands, like small needles, which were observed to be totally straight (Figure 3.10a). In contrast, BNC crumb presented

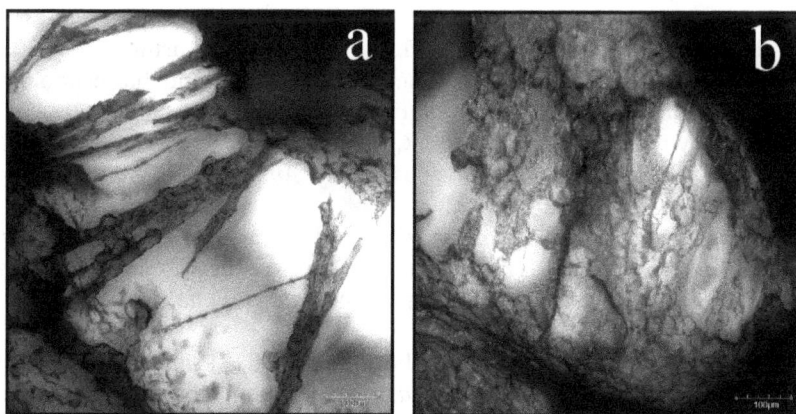

Figure 3.10 CLSM micrographs of bread crumb: a) control and b) with BNC. Black bar indicates 100 µm.

interweaved gluten filaments which though oriented, but were thicker and not completely straight (Figure 3.10b). Such thickening of gluten filaments could explain the increased gas retention of the matrix during baking, thus, producing a more porous crumb with tender texture.

3.4 Conclusions

In recent years, bacterial nanocellulose has evolved as a promising material owing to its unique properties, such as high strength, stiffness, biodegradability and renewability that open a wide spectrum of applications in different industrial and biomedical areas.

Studies on the cost-effectiveness of culture medium have provided improvements to BNC yield and productivity, giving the possibility of using agro-industrial residues to obtain BNC.

From the point of view of the food industry, even though several research studies have been carried out, there is still much work to be done. Bacterial nanocellulose is not commercially available in most of the countries, which implies a major disadvantage with respect to other hydrocolloids. However, the technical and dietetic properties reviewed in this chapter exhibit the great potential of BNC for large scale usage as a novel hydrocolloid for different applications in food technology.

References

1. Lynd, L. R., Weimer, P. J., Zyl, W. H. V., and Pretorius, I. S. (2002) Microbial cellulose utilization: fundamentals and biotechnology. *Microbiological and Molecular Biology Reviews,* **66**, 506-577.
2. Brown, Jr., R. M., Kudlicka, K., Cousins, S. K., and Nagy, R. (1992) Gravity effects on cellulose assembly. *American Journal of Botany*, **79**(11), 1247-1258.
3. Klemm, D., Kramer, F., Moritz, S., Lindström, T., Ankerfors, M., Gray, D., and Dorris, A. (2011) Nanocelluloses: a new family of nature-based materials. *Angewandte Chemie International Edition*, **50**(24), 5438-5466.
4. Strom, G., Ohgren, C., and Ankerfors, M. (2013) *Nanocellulose as an Additive in Foodstuff*, Innventia. Online: http://www.innventia.com/Documents/Rapporter/Innventia%20 report403.pdf [accessed 3rd January 2019].
5. Arola, S., Malho, J. M., Laaksonen, P., Lille, M., and Linder, M. B.

(2013) The role of hemicellulose in nanofibrillated cellulose networks. *Soft Matter*, **9**(4), 1319-1326.

6. Eronen, P., Osterberg, M., Heikkinen, S., Tenkanen, M., and Laine, J. (2011) Interactions of structurally different hemicelluloses with nanofibrillar cellulose. *Carbohydrate Polymers*, **86**(3), 1281-1290.

7. Brown, A. J. (1886) On an acetic acid ferment which forms cellulose. *Journal of the Chemical Society*, **49**, 432-439.

8. List of Prokaryotic Names with Standing in Nomenclature, LPSN (2013). Online: http://www.bacterio.net/ [accessed 4th January 2019].

9. Validation List No. 64: Validation of publication of new names and new combinations previously effectively published outside the IJSB (1998). *International Journal of Journal of Systematic Bacteriology*, **48**, 327-328.

10. Marchetti, L., Muzzio, B., Cerrutti, P., Andrés, S. C., and Califano, A. N. (2017) Impact of bacterial nanocellulose on the rheological and textural characteristics of low-lipid meat emulsions. In: *Nanotechnology Applications in Food*, Grumezescu, A. M., and Oprea, A. (eds), Elsevier, USA, pp. 345-361.

11. Valla, S., and Kjosbakken, J. (1982) Cellulose-negative mutants of *Acetobacter xylinum*. *Microbiology*, **128**(7), 1401-1408.

12. Williams, W. S., and Cannon, R. E. (1989) Alternative environmental roles for cellulose produced by *Acetobacter xylinum*. *Applied and Environmental Microbiology*, **55**(10), 2448-2452.

13. Iguchi, M., Yamanaka, S., and Budhiono, A. (2000) Bacterial cellulose a masterpiece of nature's arts. *Journal of Materials Science*, **35**(2), 261-270.

14. Pineda, L. D. C., Mesa, L. A. C., and Riascos, C. A. M. (2012) fermentation techniques and applications of bacterial cellulose: A review. *Ingeniería y Ciencia*, **16**, 307-335.

15. Ruka, D. R., Simon, G. P., and Dean, K. M. (2012) Altering the growth conditions of *Gluconacetobacter xylinus* to maximize the yield of bacterial cellulose. *Carbohydrate Polymers*, **89**(2), 613-622.

16. Foresti, M. L., Cerrutti, P., and Vázquez, A. (2015) Bacterial nanocellulose: synthesis, properties and applications. In: *Polymer Nanocomposites Based on Inorganic and Organic Nanomaterials*, Mohanty, S., Nayak, S. K., Kaith, B. S., and Kalia, S. (eds), John Wiley and Sons, USA, pp. 39-61.

17. Hestrin, S., and Schramm, M. (1954) Synthesis of cellulose by *Acetobacter xylinum*. 2. Preparation of freeze-dried cells capable of polymerizing glucose to cellulose. *Biochemical Journal*, **58**(2), 345-352.

18. Chawla, P. R., Bajaj, I. B., Survase, S. A., and Singhal, R. S. (2009) Microbial cellulose: fermentative production and applications. *Food Technology and Biotechnology*, **47**(2), 107-124.

19. Santos, S. M., Carbajo, J. M., and Villar, J. C. (2013) The effect of carbon and nitrogen sources on bacterial cellulose production and properties from *Gluconacetobacter sucrofermentans* CECT 7291 focused on its use in degraded paper restoration. *BioResources*, **8**(3), 3630-3645.
20. Castro, C., Zuluaga, R., Putaux, J. L., Caro, G., Mondragon, I., and Ganán, P. (2011) Structural characterization of bacterial cellulose produced by *Gluconacetobacter swingsii* sp. from Colombian agroindustrial wastes. *Carbohydrate Polymers*, **84**(1), 96-102.
21. Rani, M. U., Rastogi, N. K., and Appaiah, K. A. (2011) Statistical optimization of medium composition for bacterial cellulose production by *Gluconacetobacter hansenii* UAC09 using coffee cherry husk extract-an agro-industry waste. *Journal of Microbiology and Biotechnology*, **21**, 739-745.
22. Vázquez, A., Foresti, M. L., Cerrutti, P., and Galvagno, M. (2013) Bacterial cellulose from simple and low cost production media by *Gluconacetobacter xylinus*. *Journal of Polymers and the Environment*, **21**(2), 545-554.
23. Cerrutti, P., Roldán, P., García, R. M., Galvagno, M. A., Vázquez, A., and Foresti, M. L. (2016) Production of bacterial nanocellulose from wine industry residues: Importance of fermentation time on pellicle characteristics. *Journal of Applied Polymer Science*, **133**(14), 43109.
24. Klemm, D., Schumann, D., Udhardt, U., and Marsch, S. (2001) Bacterial synthesized cellulose - artificial blood vessels for microsurgery. *Progress in Polymer Science*, **26**(9), 1561-1603.
25. Klemm, D., Schumann, D., Kramer, F., Heßler, N., Hornung, M., Schmauder, H. P., and Marsch, S. (2006) Nanocelluloses as innovative polymers in research and application. In: *Polysaccharides II*, Klemm, D. (ed), Springer, Germany, pp. 49-96.
26. Zywicka, A., Peitler, D., Rakoczy, R., Konopacki, M., Kordas, M., and Fijałkowski, K. (2015) The effect of different agitation modes on bacterial cellulose synthesis by *Gluconacetobacter xylinus* strains. *Acta Scientiarum Polonorum Zootechnica*, **14**(1), 137-150.
27. Lee, K. Y., Aitomaki, Y., Berglund, L. A., Oksman, K., and Bismarck, A. (2014) On the use of nanocellulose as reinforcement in polymer matrix composites. *Composites Science and Technology*, **105**, 15-27.
28. Corral, M. L., Cerrutti, P., Vázquez, A., and Califano, A. (2017) Bacterial nanocellulose as a potential additive for wheat bread. *Food Hydrocolloids*, **67**, 189-196.
29. Gama, M., Gatenholm, P., and Klemm, D. (2013) *Bacterial Nanocellulose: A Sophisticated Multifunctional Material*, CRC Press, USA.
30. Tabuchi, M. (2007) Nanobiotech versus synthetic nanotech? *Nature Biotechnology*, **25**, 389-390.
31. Fink, H. P., Purz, H. J., Bohn, A., and Kunze, J. (1997) Investigation of

the supramolecular structure of never dried bacterial cellulose. *Macromolecular Symposia*, **120**(1), 207-217.
32. Fontana, J. D., Koop, H. S., Tiboni, M., Grzybowski, A., Pereira, A., Kruger, C. D., da Silva, M. G. R. and Wielewski, L. P. (2017) New insights on bacterial cellulose. In: *Food Biosynthesis*, Grumezescu, A. M., and Holban, A. M. (eds), Elsevier, USA, pp. 213-249.
33. Yamamoto, H., Horii, F., and Hirai, A. (1996) In situ crystallization of bacterial cellulose II. Influences of different polymeric additives on the formation of celluloses I α and I β at the early stage of incubation. *Cellulose*, 3(1), 229-242.
34. Barud, H. S., Ribeiro, C. A., Capela, J. M., Crespi, M. S., Ribeiro, S. J., and Messadeq, Y. (2011) Kinetic parameters for thermal decomposition of microcrystalline, vegetal, and bacterial cellulose. *Journal of Thermal Analysis and Calorimetry*, **105**(2), 421-426.
35. Coelho de Carvalho Benini, K. C. C., Voorwald, H. J. C., Cioffi, M. O. H., Rezende, M. C., and Arantes, V. (2018) Preparation of nanocellulose from *Imperata brasiliensis* grass using Taguchi method. *Carbohydrate Polymers*, **192**, 337-346.
36. Cheng, K. C., Catchmark, J. M., and Demirci, A. (2009) Effect of different additives on bacterial cellulose production by *Acetobacter xylinum* and analysis of material property. *Cellulose*, **16**(6), 1033.
37. Balquinta, M. L., Lorenzo, G., Andrés, S. C., and Califano, A. N. (2018) Análisis reológico y estructural de nanocelulosa bacteriana sometida a liofilización. *IV Reunión Interdisciplinaria de Tecnología y Procesos Químicos*, Argentina.
38. Kilzer, F. J., and Broido, A. (1965) Speculations on the nature of cellulose pyrolysis. *Pyrodynamics*, **2**, 151-163.
39. Oh, S. Y., Yoo, D. I., Shin, Y., Kim, H. C., Kim, H. Y., Chung, Y. S., Park, W. H., and Youk, J. H. (2005) Crystalline structure analysis of cellulose treated with sodium hydroxide and carbon dioxide by means of X-ray diffraction and FTIR spectroscopy. *Carbohydrate research*, **340**(15), 2376-2391.
40. Gea, S., Reynolds, C. T., Roohpour, N., Wirjosentono, B., Soykeabkaew, N., Bilotti, E., and Peijs, T. (2011) Investigation into the structural, morphological, mechanical and thermal behaviour of bacterial cellulose after a two-step purification process. *Bioresource Technology*, **102**(19), 9105-9110.
41. Faria Tischer, P. C., Sierakowski, M. R., Westfahl Jr, H., and Tischer, C. A. (2010) Nanostructural reorganization of bacterial cellulose by ultrasonic treatment. *Biomacromolecules*, **11**(5), 1217-1224.
42. French, A. D. (2014) Idealized powder diffraction patterns for cellulose polymorphs. *Cellulose*, **21**(2), 885-896.
43. Wada, M., and Okano, T. (2001) Localization of Iα and Iβ phases in algal cellulose revealed by acid treatments. *Cellulose*, **8**, 183-188.
44. Elazzouzi-Hafraoui, S., Nishiyama, Y., Putaux, J. L., Heux, L., Dubre-

uil, F., and Rochas, C. (2008) The shape and size distribution of crystalline nanoparticles prepared by acid hydrolysis of native cellulose. *Biomacromolecules*, **9**(1), 57-65.

45. Sugiyama, J., Vuong, R., and Chanzy, H. (1991) Electron diffraction study on the two crystalline phases occurring in native cellulose from an algal cell wall. *Macromolecules*, **24**(14), 4168-4175.

46. Fu, L., Zhou, P., Zhang, S., and Yang, G. (2013) Evaluation of bacterial nanocellulose-based uniform wound dressing for large area skin transplantation. *Materials Science and Engineering C: Materials for Biological Applications*, **33**(5), 2995-3000.

47. Amin, M. C. I. M., Abadi, A. G., and Katas, H. (2014) Purification, characterization and comparative studies of spray-dried bacterial cellulose microparticles. *Carbohydrate Polymers*, **99**, 180-189.

48. Xiang, Z., Gao, W., Chen, L., Lan, W., Zhu, J. Y., and Runge, T. (2016) A comparison of cellulose nanofibrils produced from *Cladophora glomerata* algae and bleached eucalyptus pulp. *Cellulose*, **23**(1), 493-503.

49. Dai, D., and Fan, M. (2011) Investigation of the dislocation of natural fibres by Fourier-transform infrared spectroscopy. *Vibrational Spectroscopy*, **55**(2), 300-306.

50. Sheltami, R. M., Abdullah, I., Ahmad, I., Dufresne, A., and Kargarzadeh, H. (2012) Extraction of cellulose nanocrystals from mengkuang leaves (*Pandanus tectorius*). *Carbohydrate Polymers*, **88**(2), 772-779.

51. Kargarzadeh, H., Ahmad, I., Abdullah, I., Dufresne, A., Zainudin, S. Y., and Sheltami, R. M. (2012) Effects of hydrolysis conditions on the morphology, crystallinity, and thermal stability of cellulose nanocrystals extracted from kenaf bast fibers. *Cellulose*, **19**(3), 855-866.

52. Kargarzadeh, H., Sheltami, R. M., Ahmad, I., Abdullah, I., and Dufresne, A. (2015) Cellulose nanocrystal: A promising toughening agent for unsaturated polyester nanocomposite. *Polymer*, **56**, 346-357.

53. Saelee, K., Yingkamhaeng, N., Nimchua, T., and Sukyai, P. (2016) An environmentally friendly xylanase-assisted pretreatment for cellulose nanofibrils isolation from sugarcane bagasse by high-pressure homogenization. *Industrial Crops and Products*, **82**, 149-160.

54. Shankar, S., and Rhim, J. W. (2016) Preparation of nanocellulose from micro-crystalline cellulose: The effect on the performance and properties of agar-based composite films. *Carbohydrate Polymers*, **135**, 18-26.

55. Wang, J., Rosell, C. M., and Benedito de Barbera, C.(2002) Effect of the addition of different fibres on wheat dough performance and bread quality. *Food Chemistry*, **79**, 221-226.

56. Liang, C. Y., and Marchessault, R. H. (1959) Infrared spectra of crystalline polysaccharides. II. Native celluloses in the region from 640

to 1700 cm^{-1}. *Journal of Polymer Science Part A: Polymer Chemistry*, **39**(135), 269-278.

57. Siroky, J., Blackburn, R. S., Bechtold, T., Taylor, J., and White, P. (2010) Attenuated total reflectance Fourier-transform Infrared spectroscopy analysis of crystallinity changes in lyocell following continuous treatment with sodium hydroxide. *Cellulose*, **17**(1), 103-115.

58. Schwanninger, M., Rodrigues, J. C., Pereira, H., and Hinterstoisser, B. (2004) Effects of short-time vibratory ball milling on the shape of FT-IR spectra of wood and cellulose. *Vibrational Spectroscopy*, **36**(1), 23-40.

59. Wu, J. M., and Liu, R. H. (2013) Cost-effective production of bacterial cellulose in static cultures using distillery wastewater. *Journal of Bioscience and Bioengineering*, **115**(3), 284-290.

60. Kazmierczak, D., and Kazimierczak, J. (2010) Biosynthesis of modified bacterial cellulose in a tubular form. *Fibres & Textiles in Eastern Europe*, **18**(5), 82.

61. Carrillo, F., Colom, X., Sunol, J. J., and Saurina, J. (2004) Structural FTIR analysis and thermal characterization of lyocell and viscose-type fibres. *European Polymer Journal*, **40**(9), 2229-2234.

62. Lieber, E., Rao, C. R., Chao, T. S., and Hoffman, C. W. W. (1957) Infrared spectra of organic azides. *Analytical Chemistry*, **29**(6), 916-918.

63. Zhong, Q., Steinhurst, D. A., Carpenter, E. E., and Owrutsky, J. C. (2002) Fourier transform infrared spectroscopy of azide ion in reverse micelles. *Langmuir*, **18**(20), 7401-7408.

64. *Infrared Spectroscopy Absorption Table*, Ochemonline (2018). Online: http://www.ochemonline.com/Infrared_spectroscopy_absorption_table [accessed 8th January 2019].

65. Charreau, H., L Foresti, M., and Vázquez, A. (2013) Nanocellulose patents trends: a comprehensive review on patents on cellulose nanocrystals, microfibrillated and bacterial cellulose. *Recent Patents on Nanotechnology*, **7**(1), 56-80.

66. Takai, M., Nonomura, F., Hayashi, J., Fukaya, M., and Okumura, H. (1989) Filter Membrane Containing Bacterized Cellulose, patent JP1199604A.

67. Ring, D. F., Nashed, W., and Dow, T. (1984) Foer medicinska aendamaol avsedd vaetskeimpregnerad saordyna, patent FI198304623A.

68. Iguchi, M., Mitsuhashi, S., Ichimura K, Nishi, Y., Uryu, M., Yamanaka, S., and Watanabe, K. (1986) Moulded Material Comprising Bacteria-produced Cellulose, patent EP200409A2.

69. Yamashita, S., and Takehara, M. (1989). Plaster for Oral Cavity Application, patent JP1050815A.

70. Penny, G. S., Stephens, R. S., and Winslow, A. R. (1991) Method of Supporting Fractures in Geologic Formations and Hydraulic Fluid Composition for Same, patent US5009797A.

71. Yamanaka, S., Ono, E., Watanabe, K., Kusakabe, M., and Suzuki, Y. (1990) Hollow Microbial Cellulose, Process for Preparation Thereof, and Artificial Blood Vessel Formed of Said Cellulose, patent EP396344A2.

72. Klemm, D., Marsch, S., Schumann, D., and Udhardt, U. (2001) Method and Device for Producing Shaped Microbial Cellulose for Use as Biomaterial, Especially for Microsurgery, patent WO2001061026A1.

73. Kaneko, K., and Watabe, O. (1998) Edible Bacterial Cellulose Composition, patent JP10117705A.

74. Stephens, R. S., Westland, J. A., and Neogi, A. N. (1990) Method of Using Bacterial Cellulose as a Dietary Fiber Component, patent US4960763A.

75. Shi, Z., Zhang, Y., Phillips, G.O., and Yang, G. (2014) Utilization of bacterial cellulose in food. *Food Hydrocolloids*, **35**, 539-545.

76. Dourado, F., Leal, M., Martins, D., Fontão, A., Rodrigues, A. C., and Gama, M. (2017) Celluloses as food ingredients/additives: Is there a room for BNC? In: *Bacterial Nanocellulose. From Biotechnology to Bio-Economy*, Gama, M., Dourado, F., Bielecki, S. (ed), Elsevier, Netherlands, pp. 123-133.

77. Lin, K. W., and Lin, H. Y. (2004) Quality Characteristics of Chinese-style meatball containing bacterial cellulose (Nata). *Journal of Food Science*, **69**(3), 107-111.

78. Cho, S. S., and Almeida, N. (2012) *Dietary Fiber and Health*. CRC Press, USA.

79. Purwadaria, T., Gunawan, L., Gunawan, A. W. (2010) The production on nata colored by monascus purpureus J1 pigments as functional food. *Microbiology Indonesia*, **4**, 6-10.

80. Stauffer, C. E. (1990) *Functional Additives for Bakery Foods*, Van Nostrand Reinhold, USA.

81. Wang, L. F., Shankar, S., and Rhim, J. W. (2017) Properties of alginate-based films reinforced with cellulose fibers and cellulose nanowhiskers isolated from mulberry pulp. *Food Hydrocolloids*, **63**, 201-208.

82. Eckardt, J., Ohgren, C., Alpa, A., Ekman, S., Astrom, A., Chen, G., Swenson, J., Johansson, D., and Langton, M. (2013) Long-term frozen storage of wheat bread and dough-Effect of time, temperature and fibre on sensory quality, microstructure and state of water. *Journal of Cereal Science*, **57**, 125-133.

83. Correa, M. J., Ferrer, E., Añón, M. C., and Ferrero, M. C. (2014) Interaction of modified celluloses and pectins with gluten proteins. *Food Hydrocolloids*, **35**, 91-99.

4

Nanocellulose as Reinforcement in Polymer Nanocomposites

Haleema Saleem and Vikas Mittal*,**

Department of Chemical Engineering, The Petroleum Institute (part of Khalifa University of Science and Technology), Abu Dhabi, UAE

Corresponding author: vik.mittal@gmail.com
**Current address: Bletchington, Wellington County, Australia*

4.1 Introduction

Cellulose is regarded as the most abundant organic compound generated from biomass [1]. This material has been utilized for around 150 years in numerous applications. Progressing knowledge about the reactivity and structural features of cellulose has led to the gradual development of novel cellulose-based materials. Independent of its source, cellulose has a white fiber-like structure with a specific gravity of approx. 1.5. The chemical and physical aspects of cellulose have been comprehensively studied since its discovery by Anselme Payen in 1838. It is a linear homopolysaccharide of β-1,4-linked-anhydro-D-glucose units [2,3] having a degree of polymerization (DP) of almost 20,000, but shorter cellulose chains are also observed. The fundamental chemical framework of cellulose is displayed in Figure 4.1, where a dimer known as cellobiose shows up as a repeated fragment [2].

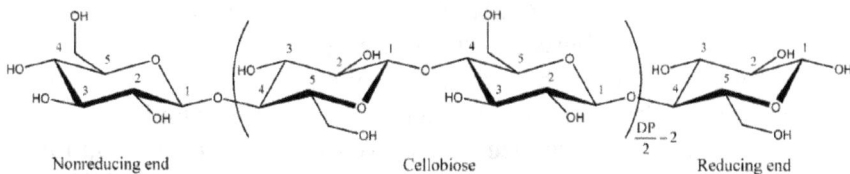

Figure 4.1 The chemical framework of cellulose. Reproduced from Reference 2 with permission from American Chemical Society.

Nanocellulose, edited by Vikas Mittal
© 2019 Central West Publishing, Australia

Cellulose has four distinct polymorphs namely cellulose I, II, III and IV. Cellulose I exists naturally and possesses two allomorphs, I_α and I_β. Cellulose II, which is the regenerated cellulose, develops subsequent to re-crystallization [4]. The primary difference between cellulose I and II lies in the arrangement of their atoms: cellulose I chains run in a side-by-side direction, whereas cellulose II has anti-parallel packing. The cellulose III_{II} and III_I forms are acquired by cellulose II and I respectively after ammonia treatment, and cellulose IV is subsequently generated by the modification of cellulose III.

The chemical modification of cellulosic materials on an industrial scale has led to the development of many advanced products, such as membranes, coatings, building materials, films, foodstuff and drilling fluids. Presently, the development of nanocellulose in the form of nanocrystals, nano-fibrils, nano-whiskers and nanofibers has gained significant research attention. Novel techniques for the generation of these nanomaterials range from top-down strategies, including physical/synthetic/enzymatic approaches for detachment from agricultural/forest residues and wood, to the bottom-up generation of cellulosic nano-fibrils from glucose by bacteria [5,6]. These materials display unique properties owing to their nanostructure such as hydrophilicity, ease of chemical modification and the generation of adaptable semi-crystalline fiber structures, along with extensive surface area.

Depending on preparation strategies, function as well as dimensions, nanocellulose can be categorized into three primary subcategories, namely micro-fibrillated cellulose (MFC), nano-crystalline cellulose (NCC) and bacterial nanocellulose (BNC) [6]. As indicated by Habibi et al. [2], around 36 distinctive cellulose molecules are united by biomass to form units termed as micro-fibrils or elementary fibrils, which are further arranged in bigger units to form micro-fibrillated cellulose [2]. The elementary fibril's diameter is almost 5 nm, while MFC, also known as nano-fibrillated cellulose (NFC), has diameter between 20 to 50 nm. A micro-fibril can be regarded as an adaptable hair strand having crystals of cellulose connected across the axis of the micro-fibril by disorganized amorphous areas [7]. The organized areas are the cellulose chain bundles which are stabilized using a powerful as well as complicated framework of hydrogen bonding [2]. NCC, otherwise called whiskers, comprises of rod-like crystals of cellulose having lengths from 100 nm to a few micrometers and widths in the range 5-70 nm. NCC is prepared by the amorphous phase elimination of a decontaminated cellulose material by acid hydrolysis. Even though comparable in size to MFC, NCC possesses exceptionally restricted flexibility and displays prolonged crystalline rod-like shapes. BNC, additionally known as bio-cellulose or microbial cellulose, is generated by the aerobic bacteria, like acetic acid bacteria of type *Gluconacetobacter* [6,8]. In contrast with the NCC and MFC materials which are segregated from the cellulose sources, BNC is generated as a polymer as well as nanomaterial by biotechnological congregation procedures from the lower molecular weight carbon sources (for example, D-glucose).

In this chapter, substantial advancements and prospects of nanocellulose-based materials are presented. Different preparation and characterization methods of the three types of nanocellulose, namely MFC, NCC and BNC, are also discussed. Finally, the chapter focuses on the potential application of nanocellulose as reinforcing agent for the generation of polymer nanocomposites.

4.2 Types of Nanocellulose

4.2.1 Micro-fibrillated Cellulose

MFC, also known as NFC, cellulose micro-fibrils or micro-fibrillar cellulose, was assessed quite recently, especially with respect to the nanocomposite applications [6,9]. The initial cellulosic material utilized to prepare MFC was wood, as detailed by Herrick *et al.* [10]. The wood pulp was broken down using a high-pressure homogenizer with an aim of acquiring a viscous as well as shear thinning aqueous gel at a lower concentration. Currently, MFC is generated from several cellulosic sources.

Synthesis of MFC

As mentioned earlier, MFC is typically generated from wood by high pressure homogenization as per the techniques advanced at ITT Rayonnier [10]. The bleached kraft pulp is generally employed as a starting material for the generation of MFC. The pulp is generated by the chemical treatment of wood by utilizing a blend of sodium sulfide and sodium hydroxide. Pulping with sulfurous acid salts gives rise to sulfite pulp. Delamination procedure has been observed to be expedited by the incorporation of hydrophilic polymers, like guar gums, carboxymethyl cellulose, poly(acrylic acid), methyl cellulose, carrageenin and hydroxyl-propyl cellulose (HPC). These polymers diminish the clogging trend and allow greater pulp consistencies at the time of homogenization. The sulfite pulps are easier to delaminate relative to the kraft pulps, and higher charge density or/and greater hemi-cellulose content assist the delamination. However, it was noted earlier that almost 27,000 kWh per ton was required for generating a gel-like MFC from a suspension of sulfite-pulp possessing a greater content of hemi-cellulose. The addition of the charged groups within the pulp fibers was observed to improve the fiber wall delamination, and by the incorporation of carboxy-methyl groups, a completely delaminated carboxy-methylated MFC might be generated [11]. The addition of the charged groups has been suggested as a genuinely effective approach to diminish the energy utilization. The use of oxidative transition metal ions to oxidize the fibers has been the subject of many patents, and, in addition, persulfate-oxidized cellulose has been reported as an appropriate substrate to prepare MFC [12]. An innovative method of attaining charged carboxylate groups in the cellulose structure together with MFC generation is the 2,2,6,6-tetramethyl-piperidine-1-oxyl (TEMPO) mediated oxidation of the 1° hydroxy groups at the cellulose C6, a technique fundamentally advanced by Saito et al. [13,14].

A wide range of delamination techniques have been developed including the utilization of micro-fluidizers [15], consolidation of beating, rubbing as well as homogenization [16], super-grinding treatments [17], cryogenic crushing in different configurations [18,19] and greater shear refining. Delamination using ball mills and ultrasonication have additionally been utilized to generate nanofibers [20]. MFC re-dispersion subsequent to drying is a challenge, as the irrevocable fibril conglomeration (a procedure termed as the hornification) brings about a material having ivory-like characteristics which make the material non-dispersible for composite applications. The principle methodology to avoid hornification is the generation of a steric hindrance or presentation of electrostatic groups for obstructing interdependent hydrogen bonding among cellulose chains. Beneficial additives used for this purpose are polyhydroxy-functionalized admixtures, especially carbohydrates or carbohydrate associated compounds, for example, starches, carbohydrate gums, seaweed extracts, glycosides, cellulose derivatives, oligosaccharides and glycol compounds [21]. Another strategy is the generation of other functional groups on cellulose, for blocking the cooperative hydrogen bonding. The carboxy-methylation process is considered to be extremely viable for restricting the cellulose fiber hornification process.

Properties of MFC

MFC displays high surface area, crystallinity and flexibility, along with possessing a large number of hydroxyl groups. The % crystallinity, DP and strength are the important physical characteristics driving the applications of MFC. Also, the degree of crystallinity is dependent on the MFC source. For instance, a greater degree of crystallinity of almost 70% and 78% was observed for soy hull and wheat straw MFC respectively [22]. Iwamoto *et al.* [23] examined the % crystallinity of MFC subsequent to number of passes via a grinder [23]. An enhancement in the number of passes (1 to 30) was observed to diminish the degree of crystallinity, which could be illustrated by CNF hornification under a greater shear rate. The other essential characteristic of MFC is its high specific area, normally examined using modeling tools. Siqueira *et al.* [24] reported the specific area of sisal MFC as 50 m^2g^{-1}, which is almost ten times higher relative to the fibers. It is often difficult to quantify this property because of the strong aggregation of MFC after the drying process. MFC also exhibits particular rheological behavior, characterized by shear thinning as well as pseudo-plasticity [10,25,26].

The thermal degradation characteristics of the MFC films have been examined to a lesser extent. In addition to the preparation procedure, the thermal performance additionally relies on the drying process (which influences the extent of hydrogen bonds). It was observed that the TEMPO-oxidized cellulose exhibited numerous degradation events [27]. The highest extent of degradation occurred during the temperature range 225-231 °C. The weight reduction enhanced remarkably with an enhancement in the sonication time as well as temperature. Quiévy *et al.* [28] also examined the impact of the drying procedure on the thermal resistance of MFC attained by homogenization. The authors analyzed 3 drying procedures: freeze-drying of frozen-cellulose suspension for 48 hours at 40 °C, oven drying for 72 hours at 80 °C and atomization (comprising of spraying of the suspension by progressing it via a nozzle and subsequently drying the wet particles using air at 200-205 °C).

NFC-based hybrids are exceedingly assuring for coatings, lightweight packaging and films [29]. Further, superior performance could be acquired by modifying the NFC hybrid frameworks. In a

study by Liimatainen *et al.* [30], talc/NFC hybrid films having a la-
mellar brick-and-mortar framework were generated. A simple vac-
uum-filtration method, followed by vacuum drying, was utilized to
manufacture the hybrid films from individual talc platelets and di-
carboxylic acid cellulose nanofibers (DCC). The DCC nanofiber's lateral dimensions extended from 3-5 nm, as demonstrated by the
atomic force micrograph (AFM) in Figure 4.2A, whereas the length
was in the micrometer scale. As indicated by the field emission
scanning electron microscopy (FESEM), the talc specimen consisted
of distinctive platelets with no aggregates (Figure 4.2B). The
talc/NFC hybrids, with talc contents from 1 to 50 wt%, were observed to be flexible with Young's modulus and tensile strength values of 12 ± 1 GPa and 211 ± 3 MPa respectively. The acquired enhancement in the mechanical properties was attributed to the effective disintegration process employed for breaking down the talc
platelet aggregates.

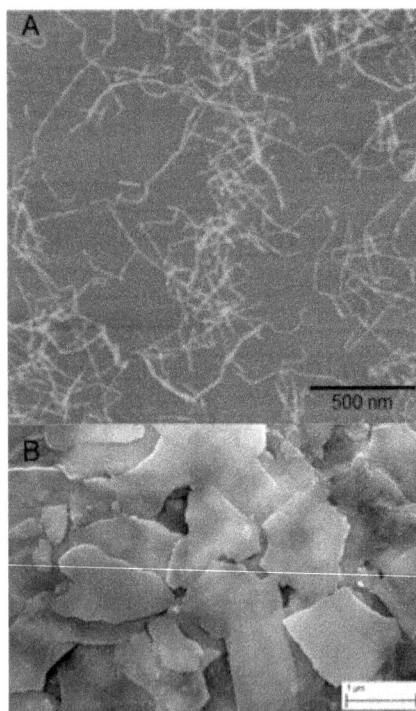

Figure 4.2 (A) AFM image of the DCC nanofibers and (B) FESEM image of
the talc platelets. Reproduced from Reference 30 with permission from
American Chemical Society.

4.2.2 Nano-crystalline Cellulose

Compared to the cellulose fibers, NCC has several benefits like greater specific strength, nanoscale dimensions, higher modulus and surface area, remarkable optical properties, etc. NCC consists of stiff rod-like crystals few hundred nanometers in length and 10-20 nm in diameter. NCC generated from native cellulose by acid hydrolysis has distinctive morphologies depending upon the hydrolysis conditions. The geometrical aspect ratio and relative degree of crystallinity are essential parameters governing the characteristics of NCC-based materials [31].

Synthesis of NCC

Acid hydrolysis is regarded as the principle procedure utilized to generate NCC. The native cellulose comprises of crystalline as well as amorphous regions, and the crystalline regions have greater density as compared to the amorphous regions. Thus, as the cellulose fibers are subjected to severe acid treatment, the amorphous sections separate, thereby, discharging the individual crystallites. The NCC properties rely upon different factors, such as temperature, reaction time, cellulose sources and types of acid utilized for the hydrolysis process.

NCC materials have been prepared using different sources, for example, cotton [32,33], wood [34], tunicate [35], sisal [36], bacterial cellulose [37], micro-crystalline cellulose [38], ramie [39] and *Valonia* cellulose [40]. NCC obtained from tunicate as well as bacterial cellulose are normally larger in dimensions as compared to cotton and wood [35,41]. This is due to the fact that the bacterial cellulose and tunicate are profoundly crystalline, thus, leading to the generation of bigger nanocrystals. Beck-Candanedo *et al.* [34] compared the characteristics of NCC from hardwood and softwood generated using similar acid-to-pulp proportions, reaction time and temperature [34]. It was noted that the NCC suspensions displayed comparative surface charge and critical concentrations (C_c). It was additionally observed that the longer reaction time and higher acid-to-pulp proportion led to smaller nanocrystals with limited polydispersity index (PDI). Due to the fact that the splitting of the cellulose chains happened arbitrarily during acid hydrolysis, the NCC dimensions were non-uniform. In another study, Bai *et al.* [42] reported a strategy for developing NCC with restricted size distribution by means of

differential centrifugation procedure [42]. Bondeson *et al.* [43] improved the reaction environment for the sulfuric acid hydrolysis of
micro-crystalline cellulose from Norway spruce, utilizing the response surface approach [43]. The concentration of sulfuric acid and
micro-crystalline cellulose, temperature, hydrolysis time and ultrasonic treatment duration were varied. It was noted that the temperature, concentration of acid and reaction time were the most important aspects for NCC generation.

For NCC generation, hydrochloric acid (HCl) and sulfuric acid
(H_2SO_4) are broadly utilized. Due to the plenitude of charged sulfate
groups existing on the surface, NCC acquired from the H_2SO_4 hydrolysis distributed promptly in H_2O as compared to the material generated from HCl hydrolysis. Also, dissimilarities in the rheological behavior and thermal stability were noted between the NCC generated
from HCl and H_2SO_4 [44]. Additionally, NCC has been prepared from
the recycled pulp utilizing the microwave supported enzymatic hydrolysis. Filson *et al.* [45] reported a technique to prepare NCC utilizing endoglucanase enzyme, a component of cellulases. The authors observed that the microwave heating resulted in higher NCC
yield as compared to normal heating.

Properties of NCC

NCC obtained from the acid hydrolysis of various forest products
has the ability to disperse in water because of negatively charged
surfaces. At lower concentration, the NCC particles are randomly
oriented as an isotropic phase in the aqueous suspensions, and as
the concentration attains a critical value, NCC generates a chiral nematic ordering [46]. With an increase in the concentration further, a
shear birefringence process is demonstrated by the aqueous suspensions of NCC. The C_c of sulfated NCC generally extends from 1 to
10% (weight/weight), which is considered to be a function of NCC
aspect ratio, osmolarity and charge density.

NCC phase behavior is responsive to the presence of electrolytes
as well as their counter particles. Impact of the incorporated electrolyte on the NCC phase segregation was analyzed by Dong *et al.*
[47]. The authors confirmed that the incorporation of the electrolytes, for example, KCl, HCl and NaCl, remarkably diminished the
anisotropic phase volume fraction. In another study, Dong *et al.* [48]
also analyzed the impact of the counter ions on the stability and
phase segregation of the NCC suspensions. For this purpose, inor-

ganic and poorly basic organic counter ions, along with immensely basic organic tetraalkyl-ammonium salts, were used. The authors also noted that the counter ion type had a remarkable impact on the phase segregation of the suspensions. Similar to the electrolytes, the phase segregation of the suspensions was firmly influenced by the inclusion of macromolecules.

Gray *et al.* [49,50] carried out a comprehensive analysis of the impact of ionic dyes and dextran on the phase equilibrium of the NCC suspensions. Heux *et al.* [41] first stated the ability of the surfactant coating to disperse the NCC whiskers in the non-polar solvents. For this purpose, the NCC whiskers from tunicate as well as cotton were blended using Beycostat NA (BNA) surfactant. Figure 4.3 shows the micrographs of cotton and tunicate suspensions in

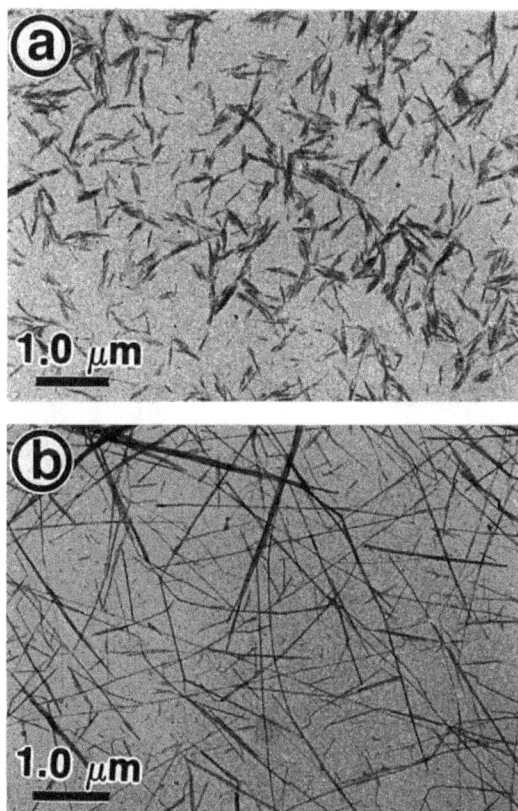

Figure 4.3 TEM micrographs of (a) cotton microcrystals and (b) tunicate microcrystals in toluene. Reproduced from Reference 41 with permission from American Chemical Society.

toluene. For the cotton suspensions (Figure 4.3a), microcrystals 8 nm in width and 200-300 nm in length were obtained. Certain aggregates of a few microcrystals, as bundles, were additionally present. In the case of tunicate suspension (Figure 4.3b) too, a uniform dispersion was observed. Cellulose microcrystals were effectively individualized, with a lateral size of approx. 15 nm and a substantial polydispersity in length, ranging from several hundreds of nanometers to a few micrometers. The authors also illustrated that the chiral nematic phases were generated despite the surfactant layer around the NCC whiskers.

In recent past, thorough analysis of the chiral nematic framework (Figure 4.4) was reported by Elazzouzi-Hafraoui *et al.* [51]. A macroscopic phase segregation with extremely clear phase boundaries as well as sharpened interfaces were noticed. Zhou *et al.* [52] also stated that the NCC material with adsorbed xyloglucan oligosaccharide-poly(ethylene glycol)-polystyrene (XGO-PEG-PS) triblock copolymer demonstrated superior dispersibility in the non-polar solvents. The quantity of the triblock copolymer adsorbed on the cellulose nanocrystals was attained from the decomposition behavior in the thermogravimetric analysis (TGA) (Figure 4.5). The thermal deterioration of primary CNC continued through two independent pyrolysis processes in the ranges of 200-290 °C and 300-400 °C, identical to spheroidal CNC having sulfate groups. Further, the inclusion of the surfactants increased the NCC dispersion in polystyrene (PS)

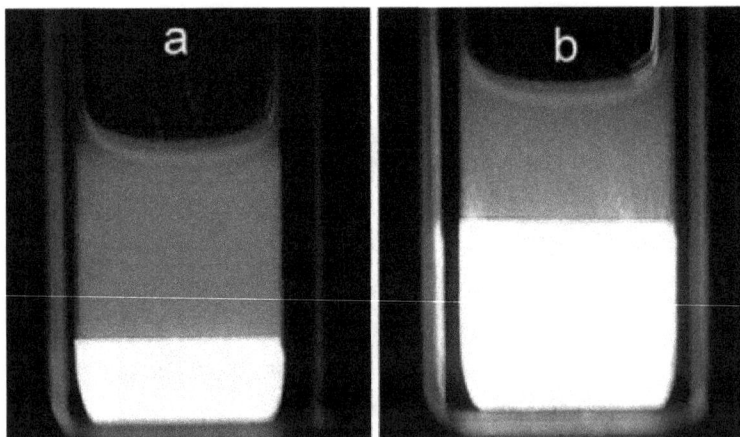

Figure 4.4 Phase separation seen among the cross-polars for varying NCC suspension concentrations of (a) 19.8 wt% and (b) 25.0 wt%. Reproduced from Reference 51 with permission from American Chemical Society.

[53]. In another study, the performance of NCC under the extrinsic field, e.g. AC electric field and magnetic field, was analyzed [54]. The impact of AC electric field on the arrangement as well as orientation of NCC was also examined by Habibi *et al.* [55]. The authors noticed that the application of the AC electric field to the NCC suspension, deposited in the middle of two metallic electrodes, led to the homogeneous arrangement of the NCC molecules. Additionally, the arrangement was significantly affected by the strength of the applied electric field as well as its frequency.

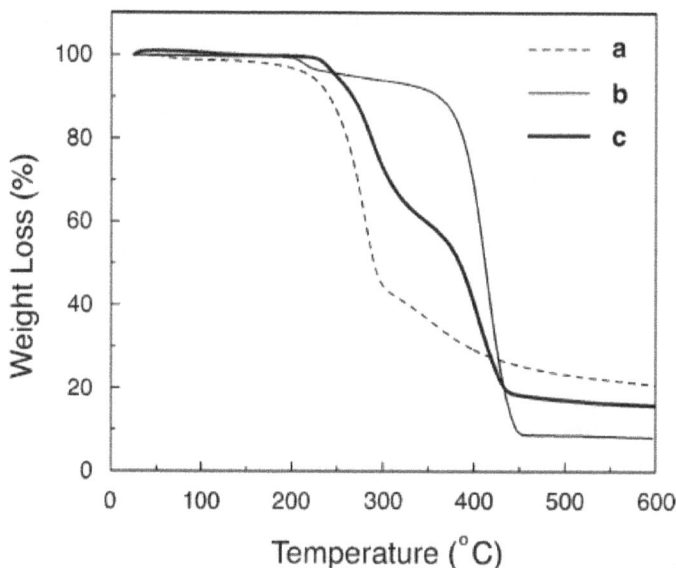

Figure 4.5 TGA curves of (a) pristine CNC, (b) XGO-PEG-PS triblock copolymer and (c) CNC adsorbed with the XGO-PEG-PS. Reproduced from Reference 52 with permission from American Chemical Society.

4.2.3 Bacterial Nanocellulose

BNC is a profoundly crystalline linear glucose polymer mostly obtained from the bacterium *Gluconacetobacter xylinus*, however, other microorganisms also display the capacity to generate BNC, e.g. other types of *Gluconacetobacter, Rhizobium* spp., *Agrobacterium tumefaciens* and gram-positive *Sarcina ventriculli* [56,57]. BNC consists of thin nanofibers with diameters ranging from 20-100 nm. It has high surface area and hydrophilic nature, which result in its excellent water absorption ability, enhanced moisture content and good adhesion [58-61]. In addition, BNC exhibits superior tensile strength, durability, biodegradability, insolubility, elasticity and non-allergenic as well as non-toxic features.

Synthesis of BNC

Acetic acid bacteria of the family *Gluconacetobacter* can generate acids as well as cellulose in high yield. The *Acetobacteraceae* are aerobic, gram-negative and rod-like microorganisms of remarkable acid tolerance, high ubiquity and dynamic motility. Remarkably, *Gluconacetobacter xylinus* cultures generate the biofilms of pristine cellulose on non-toxic surfaces [62,63]. The support material and its surface framework, *Gluconacetobacter* strain type, temperature as well as the elements of the culture medium (inclusive of additives) determine the efficiency of cellulose generation.

The bacterial cellulose preparation from the lower molecular weight sugars or alternative carbon sources by means of uridine-diphosphate glucose has been illustrated in the literature [64,65]. The generated cellulose chains discharge as fibers (diameter in na-nometer range) into the aqueous culture medium [66]. The cellulose fibrils discharged by bacteria consolidate to form ribbons and ulti-mately three dimensional (3-D) nanofiber structures.

BNC has identical molecular formula as the plant cellulose, how-ever, it is principally dissimilar due to its nanofiber structure. Sub-sequent to elementary purification, BNC contains no contaminants and functional groups, apart from the hydroxyl groups. For com-mercial utilization, cost-proficient procedures for the large scale production of BNC are needed. During the past several years, a wide range of advanced ideas have been examined. Generally, these are based on the consolidation of the agitated as well as static cultiva-tion [67-72].

Properties of BNC

Bacterial cellulose is a biomaterial with extraordinary properties, e.g. mechanical strength, chemical purity, absorbency, flexibility, size and shape compatibility, non-toxicity, inertness and bio-functionality [73]. The undried BNC has remarkable mechanical properties with stress-strain characteristics resembling with deli-cate tissue. The tensile strength of BNC pellicle has been reported to be 2 MPa in the undried state, which is considered to be high considering 99% H_2O content. In the dry-state, the stress at break of a single BNC fiber is equivalent to steel [74]. Thus, BNC is appropriate for use as a reinforcing material for polymer composites as well as paper [75,76].

4.3 Use of Nanocellulose as a Reinforcement in Polymer Nanocomposites

The initial utilization of nanocellulose as a reinforcing agent for different polymers including high density polyethylene (HDPE), PS and polypropylene (PP) was stated by Boldizar *et al.* [77]. Subsequently, Favier *et al.* [78,79] illustrated the reinforcing effect of a small fraction of nanocellulose on polymer properties. The authors utilized nanocellulose whiskers, obtained from tunicate, for reinforcing butyl acrylate and styrene copolymer latex at a loading of 6 vol%. The nanocomposites displayed superior mechanical properties as compared to the pristine polymer, which was attributed to the generation of an inflexible cellulose whisker percolation framework. Nanocellulose percolation threshold has been assessed to be in the range of 1 to 6 vol%, depending on the source of cellulose [3]. Later, Dufresne and Vignon [80] illustrated the reinforcing effect of potato pulp-extracted NFC on composite properties.

4.3.1 MFC as Reinforcement in Polymer Nanocomposites

MFC is generally utilized as a reinforcing agent for polar/hydrophilic polymers [81]. The preparation of MFC-reinforced nanocomposites with non-polar/hydrophobic polymers is restricted due to the hydrophilic characteristics of the cellulose fibers [81]. Dufresne *et al.* [82] initially illustrated the ability of MFC to enhance the mechanical properties of potato-starch [82]. Subsequently, Nakagaito *et al.* [83] demonstrated that MFC enhanced the tensile properties of a phenolic resin. In another study, Zimmermann *et al.* [84] reported an enhancement in the properties of hydroxy-propylated cellulose (HPC) on MFC incorporation. Other polymer matrices used for the development of MFC reinforced nanocomposites include polyvinyl alcohol (PVA), regenerated cellulose, poly(L-lactic acid) (PLLA) and enzymatically modified rye arabinoxylan [85-90]. Zimmermann *et al.* [84] additionally stated that the mechanical characteristics of PVA were significantly enhanced by the MFC incorporation. The Young's modulus was observed to achieve the highest value at 10 wt% fiber fraction. Lu *et al.* [86] also utilized MFC with diameter ranging from 10-100 nm and web-like framework as the reinforcement for PVA [86]. For this purpose, the authors blended a PVA aqueous solution with MFC suspensions at various proportions (1-15 wt%) and generated the composite samples using film casting. PVA crystallization, melting temperature (T_m) as well as glass transition temperature (T_g) were not affected by MFC. The authors observed 40% and 76% enhancement in modulus and strength respectively at MFC loading of 10 wt%, relative to the pristine PVA specimen. The modulus and strength did not change further on enhancing the MFC content to 15 wt%. This might be due to the accumulation of the nanosized fibers above 10% loading.

Bulota *et al.* [91] generated composite films by blending a MFC suspension with an aqueous solution of PVA using various concentrations as well as blending durations to demonstrate the impact on properties. It was observed that the mechanical properties of the composite films were dependent on the PVA-MFC mixture concentration. Further, longer blending duration negatively influenced the mechanical characteristics of the composites. Besides, the effect of the testing conditions (e.g. relative humidity) was noted to be critical because of the moisture sensitivity of cellulose fibers; higher relative humidity led to lesser strength as well as modulus.

As mentioned earlier, the utilization of hydrophobic polymers to generate MFC-filled nanocomposites is hindered by the absence of compatibility between the two phases [81]. To overcome this limitation, the fibers are required to be surface treated prior to incorporation in the polymer matrix. Suryanegara *et al.* [92] and Iwatake *et al.* [93] utilized comparable preparation strategies to generate MFC-PLA nanocomposites. Iwatake *et al.* [93] observed that the Young's modulus and tensile strength were enhanced by 40% and 25% respectively at 10 wt% MFC content, relative to the pristine polymer. Various PLA grades were utilized by Suryanegara *et al.* [92], and the authors noted small increment in the mechanical properties. Nevertheless, at the similar MFC content, Young's modulus and tensile strength enhanced by almost 35% and 10% respectively.

Lu *et al.* [81] generated hydrophobic epoxy-MFC nanocomposites, and the thermal characteristics were determined utilizing dynamic mechanical thermal analysis (DMTA). Seydibeyoğlu *et al.* [94] employed polyurethane (PU) as hydrophilic polymer matrix and determined the thermal properties utilizing DMTA. Compared to the pristine polymer [81,94], the MFC incorporation led to an enhancement in the storage modulus, E', at temperatures below or above T_g. This confirmed that the motility of the polymer chains was diminished by the inflexible MFC network and the development of interfacial bonding. Nevertheless, no variation in the T_g was observed on MFC incorporation [94]. Shimazaki *et al.* [95] also reported the development of transparent cellulose nanofiber/epoxy resin nanocomposites (Figure 4.6).

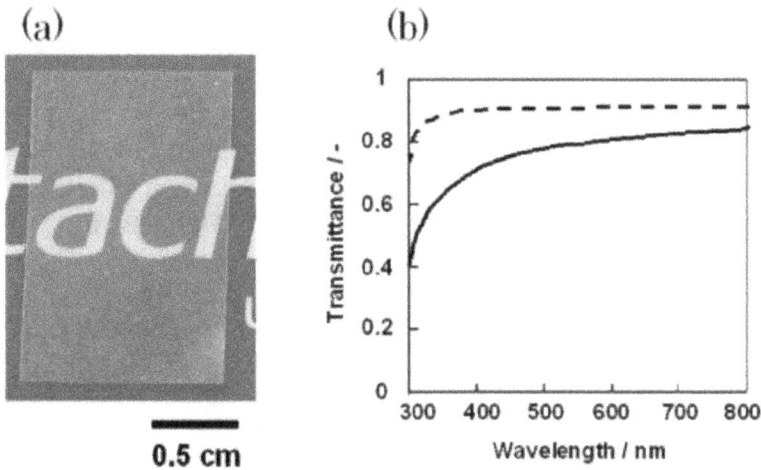

Figure 4.6 (a) Image of cellulose nanofiber/epoxy resin nanocomposite and (b) transmission spectra of nanofiber/epoxy resin nanocomposite (solid line) and epoxy resin without nanofibers (dashed line). Reproduced from Reference 95 with permission from American Chemical Society.

4.3.2 NCC as Reinforcement in Polymer Nanocomposites

NCC has a strong potential for developing lightweight, strong and inexpensive nanocomposites [2,34]. NCC reinforces the polymer matrices by the development of a percolation framework associated with the uniformly dispersed NCC phase [78,96-100].

Khan *et al.* [97] reported biodegradable films of NCC filled chitosan generated using the solution casting technique. The content of NCC varied from 1 to 10 wt%. It was noted that the nanocomposite film with 5 wt% NCC content exhibited 26% improvement in tensile strength, relative to the pure chitosan film. Further, the film demonstrated remarkably improved the barrier properties. Also, the water vapor penetrability (WVP) of the 5 wt% NCC based chitosan film diminished by 27%. In addition, the swelling analysis demonstrated a reduction in the water adsorption on the NCC filled chitosan films. The thermal properties of the polymer were not impacted by the NCC incorporation. The analysis of the surface morphology of the films revealed that the NCC phase was homogenously dispersed in the chitosan matrix.

Huq *et al.* [98] generated NCC filled alginate-based nanocomposites using the solution casting technique. The content of NCC in the nanocomposites varied from 1 to 8 wt%. At 5 wt% NCC content, the nanocomposite film demonstrated 37% higher ultimate tensile strength, relative to the control sample. The inclusion of 5 wt% NCC also reduced the nanocomposite's WVP by 31%. The FTIR analysis confirmed the molecular interaction between the NCC and alginate phases. Also, it was observed that the thermal stability of alginate-based nanocomposite films was improved subsequent to the NCC inclusion.

Li *et al.* [101] generated rod-shaped NCC from micro-crystalline cellulose (MCC) utilizing a physical strategy of high-intensity ultra-sonication. The reinforcing ability of the generated NCC was examined by incorporating it in PVA by using the solution casting technique. The diffraction analysis demonstrated that the NCC material possessed the cellulose I crystal framework. Further, it was noted that the NCC crystallinity diminished with an enhancement in the ultrasonication time. The ultrasonic impact was considered to be non-selective, which implies that it could separate the crystalline as well as amorphous cellulose. The storage modulus of NCC/PVA nanocomposites was observed to be remarkably enhanced, relative to pristine PVA. For instance, the modulus of the composite with 8 wt% NCC content was noted to be 2.4 times higher than the pristine PVA. In another study, Lalia *et al.* [102] prepared polyvinylidenefluoride-co-hexafluoropropylene (PVDF-HFP) membranes consisting of 0-4 wt% NCC using the electrospinning method. The impact of NCC on the pore size distribution, morphology, mechanical strength and liquid entry pressure (LEP) of the fibrous membranes was explored. The PVDF-HFP membrane consisting of 2 wt% NCC exhibited Young's modulus and tensile strength values of 105 MPa and 17.2 MPa respectively. The membrane demonstrated a limited pore size distribution with average pore size of 0.2 μm. The inclusion of 2 wt% NCC in the membrane also prompted a significant increment in its LEP from 19 psi to 27 psi.

4.3.3 BNC as Reinforcement in Polymer Nanocomposites

The utilization of BNC as reinforcement in nanocomposites was initially illustrated by Gindl and Keckes [103]. Cellulose acetate butyrate (CAB) matrix was reinforced using different BNC loadings. At 32 vol% BNC loading, the tensile characteristics of the CAB nanocomposite enhanced by almost 5 times, relative to the pristine polymer. It was noted that the BNC filled nanocomposite had tensile strength and modulus as high as 320 MPa and 21 GPa respectively, at BNC loading of 60 vol%. Nakagaito *et al.* [104] also reported the tensile strength of 420 MPa with 84 vol% BNC content in phenolic resin.

Various research studies have been performed to hydrophobize the BNC surface for enhancing the compatibility between the hydrophobic polymer and BNC. The hydrophobization of BNC was reported by employing dodecanoic acid in an esterification process [105]. The contact angles of the PLLA droplets on the surface of the modified BNC demonstrated enhanced wettability. The PLLA nanocomposites with modified BNC (5 wt% content) additionally indicated 12% and 50% enhancement in tensile strength and modulus respectively, relative to PLLA. Nevertheless, compared to PLLA nanocomposites with pristine BNC (5 wt% content), the improvements were marginal (only 5% improvement in tensile modulus, however, no enhancement in tensile strength). Besides, the chemical modification of BNC is considered to be extremely complex [106]. Thus, the alternative strategy of enhancing the BNC loading above 30 vol% may be more beneficial.

Stevanic *et al.* [107] illustrated the feasibility to generate xylan films with enhanced strength by the incorporation of BNC. The blending of BNC with xylan using a high-pressure micro-fluidizer resulted in optically clear nanocomposite films. Generally, the nanocomposite films were stiffer and stronger as compared to the pure polymer. Inclusion of BNC had a minor impact on the T_g of the polymer. The composites exhibited observable reduction in the moisture sensitivity and enhanced stiffness. In another study, Phomrak *et al.* [108] reinforced natural rubber (NR) with BNC and developed the composite films by means of a latex aqueous micro-dispersion technique. The BNC nanofibers were observed to be uniformly distributed and sufficiently dispersed in the NR matrix. It was observed that the crystallinity, opacity and hydrophilicity of the BNC/NR composite films enhanced with an increment in the BNC content. The composite with 80 wt% BNC, a hard as well as solid bio-composite film, demonstrated the Young's modulus and tensile strength of 4,128 MPa and 75 MPa respectively. Further, the 20 wt% BNC filled NR film showed an elastic elongation of 14.6% with relative enhancements in thermal stability, tensile strength and Young's modulus, relative to pristine NR.

4.4 Summary

In this chapter, the synthesis and properties of different nanocellulose types have been briefly reviewed. Subsequently, various studies reporting the use of these nanocellulose types as reinforcement to generate polymer nanocomposites have been summarized.

References

1. Lavoine, N., Desloges, I., Dufresne, A., and Bras, J. (2012) Microfibrillated cellulose–Its barrier properties and applications in cellulosic materials: A review. *Carbohydrate Polymers*, **90**(2), 735-764.
2. Habibi, Y., Lucia, L. A., and Rojas, O. J. (2010) Cellulose nanocrystals: chemistry, self-assembly, and applications. *Chemical Reviews*, **110**(6), 3479-3500.
3. Siqueira, G., Bras, J., and Dufresne, A. (2010) Cellulosic bionanocomposites: a review of preparation, properties and applications. *Polymers*, **2**(4), 728-765.
4. Aulin, C., Ahola, S., Josefsson, P., Nishino, T., Hirose, Y., Österberg, M., and Wågberg, L. (2009) Nanoscale cellulose films with different crystallinities and mesostructures: Their surface properties and interaction with water. *Langmuir*, **25**(13), 7675-7685.
5. Moon, R. J., Martini, A., Nairn, J., Simonsen, J., and Youngblood, J. (2011) Cellulose nanomaterials review: Structure, properties and nano-composites. *Chemical Society Reviews*, **40**(7), 3941-3994.
6. Klemm, D., Kramer, F., Moritz, S., Lindström, T., Ankerfors, M., Gray, D., and Dorris, A. (2011) Nano-celluloses: A new family of nature-based materials. *Angewandte Chemie International Edition*, **50**(24), 5438-5466.
7. Azizi Samir, M. A. S., Alloin, F., and Dufresne, A. (2005) Review of recent research into cellulosic whiskers, their properties and their application in nanocomposite field. *Biomacromolecules*, **6**(2), 612-626.
8. Klemm, D., Schumann, D., Kramer, F., Heßler, N., Koth, D., and Sultanova, B. (2009) Nanocellulose materials - Different cellulose, different functionality. *Macromolecular Symposia*, **280**(1), 60-71.
9. Siró, I., and Plackett, D. (2010) Microfibrillated cellulose and new

nanocomposite materials: a review. *Cellulose*, **17**(3), 459-494.
10. Herrick, F.W. Casebier, R. L. Hamilton, J. K. Sandberg, K. R. (1983) Microfibrillated cellulose: Morphology and accessibility. *Journal of Applied Polymer Science, Applied Polymer Sympoisum,* **37**, 797-813.
11. Wågberg, L., Winter, L., Ödberg, L., and Lindström, T. (1987) On the charge stoichiometry upon adsorption of a cationic polyelectrolyte on cellulosic materials. *Colloids and Surfaces,* **27**(1-3), 163-173.
12. Banker, G. S., and Kumar, V. (1995) Microfibrillated Oxycellulose, patent US5405953.
13. Saito, T., Nishiyama, Y., Putaux, J. L., Vignon, M., and Isogai, A. (2006) Homogeneous suspensions of individualized microfibrils from TEMPO-catalyzed oxidation of native cellulose. *Biomacromolecules,* **7**(6), 1687-1691.
14. Saito, T., and Isogai, A. (2007) Preparation of cellulose single microfibrils from native celluloses by TEMPO-mediated oxidation. *Cellulose Communications,* **14**(2), 62-66.
15. Henriksson, M., Henriksson, G., Berglund, L. A., and Lindström, T. (2007) An environmentally friendly method for enzyme-assisted preparation of microfibrillated cellulose (MFC) nanofibers. *European Polymer Journal,* **43**(8), 3434-3441.
16. Matsuda, Y. (2000) Properties and use of microfibrillated cellulose as papermaking addition. *Sen-i Gakkaishi,* **56**(7), 192-196.
17. Iwamoto, S., Nakagaito, A. N., Yano, H., and Nogi, M. (2005) Optically transparent composites reinforced with plant fiber-based nanofibers. *Applied Physics A: Materials Science and Processing,* **81**(6), 1109-1112.
18. Janardhnan, S., and Sain, M. M. (2007) Isolation of cellulose microfibrils–an enzymatic approach. *Bioresources,* **1**(2), 176-188.
19. Wang, B., and Sain, M. (2007) Dispersion of soybean stock-based nanofiber in a plastic matrix. *Polymer International,* **56**(4), 538-546.
20. Zhao, H. P., Feng, X. Q., and Gao, H. (2007) Ultrasonic technique for extracting nanofibers from nature materials. *Applied Physics Letters,* **90**(7), 073112.
21. Lowys, M. P., Desbrieres, J., and Rinaudo, M. (2001) Rheological characterization of cellulosic microfibril suspensions. Role of polymeric additives. *Food Hydrocolloids,* **15**(1), 25-32.
22. Alemdar, A., and Sain, M. (2008) Isolation and characterization of nanofibers from agricultural residues – Wheat straw and soy hulls. *Bioresource Technology,* **99**(6), 1664-1671.
23. Iwamoto, S., Nakagaito, A. N., and Yano, H. (2007) Nano-fibrillation of pulp fibers for the processing of transparent nano-composites. *Applied Physics A: Materials Science and Processing,* **89**(2), 461-466.
24. Siqueira, G., Bras, J., and Dufresne, A. (2009) New process of chem-

ical grafting of cellulose nanoparticles with a long chain isocyanate. *Langmuir*, **26**(1), 402-411.

25. Pääkkö, M., Ankerfors, M., Kosonen, H., Nykänen, A., Ahola, S.,
    Österberg, M., Ruokolainen, J., Laine, J., Larsson, P.T., Ikkala, O. and
    Lindström, T. (2007) Enzymatic hydrolysis combined with me-
    chanical shearing and high-pressure homogenization for nanoscale
    cellulose fibrils and strong gels. *Biomacromolecules*, **8**(6), 1934-
    1941.

26. Goussé, C., Chanzy, H., Cerrada, M. L., and Fleury, E. (2004) Surface
    silylation of cellulose microfibrils: preparation and rheological
    properties. *Polymer*, **45**(5), 1569-1575.

27. Johnson, R. K., Zink-Sharp, A., Renneckar, S. H., and Glasser, W. G.
    (2009) A new bio-based nanocomposite: fibrillated TEMPO-
    oxidized celluloses in hydroxypropylcellulose matrix. *Cellulose*,
    16(2), 227-238.

28. Quiévy, N., Jacquet, N., Sclavons, M., Deroanne, C., Paquot, M., and
    Devaux, J. (2010) Influence of homogenization and drying on the
    thermal stability of microfibrillated cellulose. *Polymer Degradation
    and Stability*, **95**(3), 306-314.

29. Ridgway, C. J., and Gane, P. A. (2012) Constructing NFC-pigment
    composite surface treatment for enhanced paper stiffness and sur-
    face properties. *Cellulose*, **19**(2), 547-560.

30. Liimatainen, H., Ezekiel, N., Sliz, R., Ohenoja, K., Sirviö, J. A., Ber-
    glund, L., Hormi, O and Niinimäki, J. (2013) High-strength nanocel-
    lulose–talc hybrid barrier films. *ACS Applied Materials & Interfaces*,
    5(24), 13412-13418.

31. Peng, B. L., Dhar, N., Liu, H. L., and Tam, K. C. (2011) Chemistry and
    applications of nanocrystalline cellulose and its derivatives: a nan-
    otechnology perspective. *The Canadian Journal of Chemical Engi-
    neering*, **89**(5), 1191-1206.

32. Araki, J., Wada, M., and Kuga, S. (2001) Steric stabilization of a cel-
    lulose microcrystal suspension by poly (ethylene glycol) grafting.
    Langmuir, **17**(1), 21-27.

33. Miller, A. F., and Donald, A. M. (2003) Imaging of anisotropic cellu-
    lose suspensions using environmental scanning electron micros-
    copy. *Biomacromolecules*, **4**(3), 510-517.

34. Beck-Candanedo, S., Roman, M., and Gray, D. G. (2005) Effect of re-
    action conditions on the properties and behavior of wood cellulose
    nanocrystal suspensions. *Biomacromolecules*, **6**(2), 1048-1054.

35. de Souza Lima, M. M., Wong, J. T., Paillet, M., Borsali, R., and Pecora,
    R. (2003) Translational and rotational dynamics of rodlike cellu-
    lose whiskers. *Langmuir*, **19**(1), 24-29.

36. de Rodriguez, N. L. G., Thielemans, W., and Dufresne, A. (2006) Si-
    sal cellulose whiskers reinforced polyvinyl acetate nano-
    composites. *Cellulose*, **13**(3), 261-270.

37. Roman, M., and Winter, W. T. (2004) Effect of sulfate groups from sulfuric acid hydrolysis on the thermal degradation behavior of bacterial cellulose. *Biomacromolecules*, **5**(5), 1671-1677.
38. Pranger, L., and Tannenbaum, R. (2008) Biobased nano-composites prepared by in situ polymerization of furfuryl alcohol with cellulose whiskers or montmorillonite clay. *Macromolecules*, **41**(22), 8682-8687.
39. de Menezes, A. J., Siqueira, G., Curvelo, A. A., and Dufresne, A. (2009) Extrusion and characterization of functionalized cellulose whiskers reinforced polyethylene nano-composites. *Polymer*, **50**(19), 4552-4563.
40. Revol, J. F. (1982) On the cross-sectional shape of cellulose crystallites in Valonia ventricosa. *Carbohydrate Polymers*, **2**(2), 123-134.
41. Heux, L., Chauve, G., and Bonini, C. (2000) Nonflocculating and chiral-nematic self-ordering of cellulose microcrystals suspensions in nonpolar solvents. *Langmuir*, **16**(21), 8210-8212.
42. Bai, W., Holbery, J., and Li, K. (2009) A technique for production of nanocrystalline cellulose with a narrow size distribution. *Cellulose*, **16**(3), 455-465.
43. Bondeson, D., Mathew, A., and Oksman, K. (2006) Optimization of the isolation of nanocrystals from microcrystalline cellulose by acid hydrolysis. *Cellulose*, **13**(2), 171-180.
44. Araki, J., Wada, M., Kuga, S., and Okano, T. (1998) Flow properties of microcrystalline cellulose suspension prepared by acid treatment of native cellulose. *Colloids and Surfaces A: Physicochemical and Engineering Aspects*, **142**(1), 75-82.
45. Filson, P. B., Dawson-Andoh, B. E., and Schwegler-Berry, D. (2009) Enzymatic-mediated production of cellulose nanocrystals from recycled pulp. *Green Chemistry*, **11**(11), 1808-1814.
46. Revol, J. F., Bradford, H., Giasson, J., Marchessault, R. H., and Gray, D. G. (1992) Helicoidal self-ordering of cellulose microfibrils in aqueous suspension. *International Journal Of Biological Macromolecules*, **14**(3), 170-172.
47. Dong, X. M., Kimura, T., Revol, J. F., and Gray, D. G. (1996) Effects of ionic strength on the isotropic– chiral nematic phase transition of suspensions of cellulose crystallites. *Langmuir*, **12**(8), 2076-2082.
48. Dong, X. M., and Gray, D. G. (1997) Effect of counterions on ordered phase formation in suspensions of charged rod-like cellulose crystallites. *Langmuir*, **13**(8), 2404-2409.
49. Edgar, C. D., and Gray, D. G. (2002) Influence of dextran on the phase behavior of suspensions of cellulose nanocrystals. *Macromolecules*, **35**(19), 7400-7406.
50. Beck-Candanedo, S., Viet, D., and Gray, D. G. (2006) Induced phase separation in cellulose nanocrystal suspensions containing ionic dye species. *Cellulose*, **13**(6), 629-635.

51. Elazzouzi-Hafraoui, S., Putaux, J. L., and Heux, L. (2009) Self-assembling and chiral nematic properties of organophilic cellulose nanocrystals. *The Journal of Physical Chemistry B*, **113**(32), 11069-11075.
52. Zhou, Q., Brumer, H., and Teeri, T. T. (2009) Self-organization of cellulose nanocrystals adsorbed with xyloglucan oligosaccharide-poly(ethylene glycol)-polystyrene triblock copolymer. *Macromolecules*, **42**(15), 5430-5432.
53. Kim, J., Montero, G., Habibi, Y., Hinestroza, J. P., Genzer, J., Argyropoulos, D. S., and Rojas, O. J. (2009) Dispersion of cellulose crystallites by nonionic surfactants in a hydrophobic polymer matrix. *Polymer Engineering and Science*, **49**(10), 2054-2061.
54. Sugiyama, J., Chanzy, H., and Maret, G. (1992) Orientation of cellulose microcrystals by strong magnetic fields. *Macromolecules*, **25**(16), 4232-4234.
55. Habibi, Y., Heim, T., and Douillard, R. (2008) AC electric field-assisted assembly and alignment of cellulose nanocrystals. *Journal of Polymer Science, Part B: Polymer Physics*, **46**(14), 1430-1436.
56. Mohammadkazemi, F., Azin, M., and Ashori, A. (2015) Production of bacterial cellulose using different carbon sources and culture media. *Carbohydrate Polymers*, **117**, 518-523.
57. Tanskul, S., Amornthatree, K., and Jaturonlak, N. (2013) A new cellulose-producing bacterium, Rhodococcus sp. MI 2: Screening and optimization of culture conditions. *Carbohydrate polymers*, **92**(1), 421-428.
58. Fu, L., Zhang, J., and Yang, G. (2013) Present status and applications of bacterial cellulose-based materials for skin tissue repair. *Carbohydrate Polymers*, **92**(2), 1432-1442.
59. Numata, Y., Sakata, T., Furukawa, H., and Tajima, K. (2015) Bacterial cellulose gels with high mechanical strength. *Materials Science and Engineering C*, **47**, 57-62.
60. Keshk, S. M. (2014) Bacterial cellulose production and its industrial applications. *Journal of Bioprocessing and Biotechniques*, **4**(2), doi:10.4172/2155-9821.1000150
61. Kurosumi, A., Sasaki, C., Yamashita, Y., and Nakamura, Y. (2009) Utilization of various fruit juices as carbon source for production of bacterial cellulose by Acetobacter xylinum NBRC 13693. *Carbohydrate Polymers*, **76**(2), 333-335.
62. Klemm, D., Heublein, B., Fink, H. P., and Bohn, A. (2005) Cellulose: faszinierendes Biopolymer und nachhaltiger Rohstoff. *Angewandte Chemie*, **117**(22), 3422-3458.
63. Brown, A. J. (1886) XIX. The chemical action of pure cultivations of bacterium aceti. *Journal of the Chemical Society, Transactions*, **49**, 172-187.

64. Yamanaka, S., Watanabe, K., Kitamura, N., Iguchi, M., Mitsuhashi, S., Nishi, Y., and Uryu, M. (1989) The structure and mechanical properties of sheets prepared from bacterial cellulose. *Journal of Materials Science*, **24**(9), 3141-3145.

65. Salmon, S., and Hudson, S. M. (1997) Crystal morphology, biosynthesis, and physical assembly of cellulose, chitin, and chitosan. *Journal of Macromolecular Science, Part C: Polymer Reviews*, **37**(2), 199-276.

66. Jonas, R., and Farah, L. F. (1998) Production and application of microbial cellulose. *Polymer Degradation and Stability*, **59**(1-3), 101-106.

67. Chao, Y. P., Sugano, Y., Kouda, T., Yoshinaga, F., and Shoda, M. (1997) Production of bacterial cellulose by Acetobacter xylinumwith an air-lift reactor. *Biotechnology Techniques*, **11**(11), 829-832.

68. Sakairi, N., Asano, H., Ogawa, M., Nishi, N., and Tokura, S. (1998) A method for direct harvest of bacterial cellulose filaments during continuous cultivation of Acetobacter xylinum. *Carbohydrate Polymers*, **35**(3-4), 233-237.

69. Bungay, H. R., and Serafica, G. C. (2000) Production of Microbial Cellulose, patent US6071727.

70. Chawla, P. R., Bajaj, I. B., Survase, S. A., and Singhal, R. S. (2009) Microbial cellulose: fermentative production and applications. *Food Technology and Biotechnology*, **47**(2), 107-124.

71. Dahman, Y. (2009) Nanostructured biomaterials and biocomposites from bacterial cellulose nanofibers. *Journal of Nanoscience and Nanotechnology*, **9**(9), 5105-5122.

72. Gardner, D. J., Oporto, G. S., Mills, R., and Samir, M. A. S. A. (2008) Adhesion and surface issues in cellulose and nanocellulose. *Journal of Adhesion Science and Technology*, **22**(5-6), 545-567.

73. Dinand, E., Chanzy, H., and Vignon, M. R. (1996) Parenchymal cell cellulose from sugar beet pulp: preparation and properties. *Cellulose*, **3**(1), 183-188.

74. Yano, H., Sugiyama, J., Nakagaito, A. N., Nogi, M., Matsuura, T., Hikita, M., and Handa, K. (2005) Optically transparent composites reinforced with networks of bacterial nanofibers. *Advanced Materials*, **17**(2), 153-155.

75. Yano, H., and Nakahara, S. (2004) Bio-composites produced from plant microfiber bundles with a nanometer unit web-like network. *Journal of Materials Science*, **39**(5), 1635-1638.

76. O'sullivan, A. C. (1997) Cellulose: the structure slowly unravels. *Cellulose*, **4**(3), 173-207.

77. Boldizar, A., Klason, C., Kubat, J., Näslund, P., and Saha, P. (1987) Prehydrolyzed cellulose as reinforcing filler for thermoplastics. *International Journal of Polymeric Materials*, **11**(4), 229-262.

78. Favier, V., Chanzy, H., and Cavaille, J. Y. (1995) Polymer nano-composites reinforced by cellulose whiskers. *Macromolecules*, **28**(18), 6365-6367.

79. Favier, V., Canova, G. R., Cavaillé, J. Y., Chanzy, H., Dufresne, A., and Gauthier, C. (1995) Nanocomposite materials from latex and cellulose whiskers. *Polymers for Advanced Technologies*, **6**(5), 351-355.

80. Dufresne, A., and Vignon, M. R. (1998) Improvement of starch film performances using cellulose microfibrils. *Macromolecules*, **31**(8), 2693-2696.

81. Lu, J., Askeland, P., and Drzal, L. T. (2008) Surface modification of microfibrillated cellulose for epoxy composite applications. *Polymer*, **49**(5), 1285-1296.

82. Dufresne, A., Dupeyre, D., and Vignon, M. R. (2000) Cellulose microfibrils from potato tuber cells: processing and characterization of starch–cellulose microfibril composites. *Journal of Applied Polymer Science*, **76**(14), 2080-2092.

83. Nakagaito, A. N., and Yano, H. (2004) The effect of morphological changes from pulp fiber towards nano-scale fibrillated cellulose on the mechanical properties of high-strength plant fiber based composites. *Applied Physics A*, **78**(4), 547-552.

84. Zimmermann, T., Pöhler, E., and Geiger, T. (2004) Cellulose fibrils for polymer reinforcement. *Advanced Engineering Materials*, **6**(9), 754-761.

85. Hornung, M., Ludwig, M., and Schmauder, H. P. (2007) Optimizing the production of bacterial cellulose in surface culture: A novel aerosol bioreactor working on a fed batch principle (part 3). *Engineering in Life Sciences*, **7**(1), 35-41.

86. Lu, J., Wang, T., and Drzal, L. T. (2008) Preparation and properties of microfibrillated cellulose polyvinyl alcohol composite materials. *Composites Part A: Applied Science and Manufacturing*, **39**(5), 738-746.

87. Pullawan, T., Wilkinson, A. N., and Eichhorn, S. J. (2010) Discrimination of matrix–fibre interactions in all-cellulose nano-composites. *Composites Science and Technology*, **70**(16), 2325-2330.

88. Jonoobi, M., Harun, J., Mathew, A. P., and Oksman, K. (2010) Mechanical properties of cellulose nanofiber (CNF) reinforced polylactic acid (PLA) prepared by twin screw extrusion. *Composites Science and Technology*, **70**(12), 1742-1747.

89. Mikkonen, K. S., Pitkänen, L., Liljeström, V., Bergström, E. M., Serimaa, R., Salmén, L., and Tenkanen, M. (2012) Arabinoxylan structure affects the reinforcement of films by microfibrillated cellulose. *Cellulose*, **19**(2), 467-480.

90. Kroon-Batenburg, L. M. J., Kroon, J., and Northolt, M. G. (1986) Chain modulus and intramolecular hydrogen bonding in native

and regenerated cellulose fibers. *Polymer Communications*, **27**(10), 290-292.

91. Bulota, M., Jääskeläinen, A. S., Paltakari, J., and Hughes, M. (2011) Properties of biocomposites: influence of preparation method, testing environment and a comparison with theoretical models. *Journal of Materials Science*, **46**(10), 3387-3398.

92. Suryanegara, L., Nakagaito, A. N., and Yano, H. (2009) The effect of crystallization of PLA on the thermal and mechanical properties of microfibrillated cellulose-reinforced PLA composites. *Composites Science and Technology*, **69**(7), 1187-1192.

93. Iwatake, A., Nogi, M., and Yano, H. (2008) Cellulose nanofiber-reinforced polylactic acid. *Composites Science and Technology*, **68**(9), 2103-2106.

94. Seydibeyoğlu, M. Ö., and Oksman, K. (2008) Novel nano-composites based on polyurethane and micro fibrillated cellulose. *Composites Science and Technology*, **68**(3), 908-914.

95. Shimazaki, Y., Miyazaki, Y., Takezawa, Y., Nogi, M., Abe, K., Ifuku, S., and Yano, H. (2007) Excellent thermal conductivity of transparent cellulose nanofiber/epoxy resin nano-composites. *Biomacromolecules*, **8**(9), 2976-2978.

96. Cao, X., Chen, Y., Chang, P. R., Muir, A. D., and Falk, G. (2008) Starch-based nano-composites reinforced with flax cellulose nanocrystals. *Express Polymer Letters*, **2**(7), 502-510.

97. Khan, A., Khan, R. A., Salmieri, S., Le Tien, C., Riedl, B., Bouchard, J., Chauve, G., Tan, V., Kamal, M.R. and Lacroix, M. (2012) Mechanical and barrier properties of nanocrystalline cellulose reinforced chitosan based nanocomposite films. *Carbohydrate Polymers*, **90**(4), 1601-1608.

98. Huq, T., Salmieri, S., Khan, A., Khan, R. A., Le Tien, C., Riedl, B., Fraschini, C., Bouchard, J., Uribe-Calderon, J., Kamal, M. R. and Lacroix, M. (2012) Nanocrystalline cellulose (NCC) reinforced alginate based biodegradable nanocomposite film. *Carbohydrate Polymers*, **90**(4), 1757-1763.

99. Lange, J., and Wyser, Y. (2003) Recent innovations in barrier technologies for plastic packaging - A review. *Packaging Technology and Science*, **16**(4), 149-158.

100. Wan, W. K., Campbell, G., Zhang, Z. F., Hui, A. J., and Boughner, D. R. (2002) Optimizing the tensile properties of polyvinyl alcohol hydrogel for the construction of a bioprosthetic heart valve stent. *Journal of Biomedical Materials Research Part A*, **63**(6), 854-861.

101. Li, W., Yue, J., and Liu, S. (2012) Preparation of nanocrystalline cellulose via ultrasound and its reinforcement capability for poly (vinyl alcohol) composites. *Ultrasonics Sonochemistry*, **19**(3), 479-485.

102. Lalia, B. S., Guillen, E., Arafat, H. A., and Hashaikeh, R. (2014) Nano-

crystalline cellulose reinforced PVDF-HFP membranes for membrane distillation application. *Desalination*, **332**(1), 134-141.

103. Gindl, W., and Keckes, J. (2004) Tensile properties of cellulose acetate butyrate composites reinforced with bacterial cellulose. *Composites Science and Technology*, **64**(15), 2407-2413.

104. Nakagaito, A. N., Iwamoto, S., and Yano, H. (2005) Bacterial cellulose: the ultimate nano-scalar cellulose morphology for the production of high-strength composites. *Applied Physics A*, **80**(1), 93-97.

105. Lee, K. Y., Blaker, J. J., and Bismarck, A. (2009) Surface functionalisation of bacterial cellulose as the route to produce green polylactide nano-composites with improved properties. *Composites Science and Technology*, **69**(15), 2724-2733.

106. Lee, K. Y., and Bismarck, A. (2012) Susceptibility of never-dried and freeze-dried bacterial cellulose towards esterification with organic acid. *Cellulose*, **19**(3), 891-900.

107. Stevanic, J. S., Joly, C., Mikkonen, K. S., Pirkkalainen, K., Serimaa, R., Rémond, C., Toriz, G., Gatenholm, P., Tenkanen, M., and Salmén, L. (2011) Bacterial nanocellulose-reinforced arabinoxylan films. *Journal of Applied Polymer Science*, **122**(2), 1030-1039.

108. Phomrak, S., and Phisalaphong, M. (2017) Reinforcement of natural rubber with bacterial cellulose via a latex aqueous microdispersion process. *Journal of Nanomaterials*, **2017**, Article ID 4739793.

5

Nanocellulose Produced from Rice Hulls

Janaina Mantovan,[a] Gizilene Maria de Carvalho[b] and Suzana Mali[a]

[a]Department of Biochemistry and Biotechnology, CCE, State University of Londrina, PO BOX 6001, 86051-990, Londrina - PR, Brazil
[b]Department of Chemistry, CCE, State University of Londrina, Londrina - PR, Brazil

*Corresponding author: smali@uel.br

5.1 Introduction

The world's population is expected to reach 9 billion people in 2050, and the use of more sustainable and environmentally friendly materials is an issue attracting the increased attention of researchers worldwide. The high production rate along with the low feedstock cost of lignocellulosic residues classify them as promising raw materials for value added product development [1,2].

According to Islam *et al.* [3], the annual global production of lignocellulosic biomass is approximately 220 billion tons on dry mass basis. This material finds use as feed for animals, is burnt in open environment or remains unutilized. Lignocellulosic residues generated from agriculture and industrial processes are mainly composed of cellulose, hemicellulose and lignin. Several applications have been proposed for this group of materials, such as bio-oil, bio-fuel, bacterial cellulose, bioethanol and other biomolecular production. In the last several years, these materials have garnered attention as an attractive source for the extraction of low-cost and high-performance cellulose and nanocellulose due to their specific features, such as abundant availability and renewable source [4,5].

Rice hulls are a major residue from rice milling industry and are considered as the most available cellulose source in the world [6]. Rice hulls are mainly composed of cellulose, but the main challenge hindering their use is the high silica content. This chapter discusses the use of rice hulls as a raw material to obtain cellulose and nanocellulose, the challenges and advantages of using this lignocellulosic residue, and the protocols reported in the literature studies for this purpose.

5.2 Rice Hulls

According to the United States Department of Agriculture, the world production of milled rice during the 2016/2017 harvest was estimated at 486 million tons, and the world's consumption was estimated at 480 million tons [7], making rice the second most produced staple crop in the world [8]. Rice is consumed on all continents and has a strong economic and social role for nations in Asia and Africa. Overall, it is considered as a staple food for approximately 50% of the world's population. China and India alone accounted for over 51% of both the global production and consumption during this period [9,10]. Brazil is the major rice producer in South America, with 12 million tons being produced during the 2016/2017 harvest [7].

Rice hulls are the hard protective coverings of rice grains and are rich in lignin and silica, rendering them indigestible for humans. The hulls are removed in the first stage of the milling process, resulting in brown rice, which is later milled again to remove the bran layer to result in white rice [11]. Quispe *et al.* [12] calculated that approximately 134 million tons of rice hulls were available worldwide annually and that a major fraction is burned in the open air or discharged into rivers and oceans, resulting in a negative impact on the environment.

Rice hulls represent approximately 20% of dried paddies [14-16], and the chemical composition of rice hulls can vary depending on the type of paddy, climatic and geographical conditions, sample preparation, method of analysis, etc. [16]. Rice hulls contain approximately 29-33% cellulose, 13-33% hemicellulose, 7-27% lignin and 13-25% silica [3,14,17-20].

Rice hulls are considered inappropriate as feed for animals due to high silica and low protein contents [18,21]. The natural degradation of rice hulls is affected negatively by their irregular surface and high silica content, making them a potential candidate for environmental pollution. Additionally, burning rice hulls results in an ash material in which the total silica content can exceed 90% [21].

Figure 5.1 shows the outer surface morphology of a rice hull. The original fiber forms a well-organized and compact structure typical of lignocellulosic materials composed of cellulose, hemicellulose and lignin, which are intermeshed and chemically bonded by non-covalent forces and covalent cross linkages [22]. Silica is found to be deposited throughout its outer surface, commonly observed in literature studies [23,24]. The inner surface morphology of rice hull can be observed in Figure 5.2, which presents a smooth surface arranged as linear ridges, which was has also been reported by other authors [23]. Ludueña *et al.* [23] reported that the outer surface of rice hulls is roughened, and silica appears to be present throughout the outer surface of this material, mainly located in the tip of domes. The surface composition of raw rice hulls determined by energy dispersion spectrum (EDS) resulted in a silica content of 19%, which is in agreement with other literature reports [6,11,23]. In another study, Islam *et al.* [3] reported a silica content of 13% for raw rice hulls.

Figure 5.1 Micrograph of outer surface of raw rice hull obtained by scanning electron microscopy (SEM).

Rice hulls have been studied for several applications, such as biosorption for heavy metal and dye removal in industrial wastewater [11,25-31], bioethanol production [20, 32-34] or energy generation [12, 35-39]. According to Bazargan *et al.* [40], the main disadvantage of using rice hulls for electricity generation is the high ash content, resulting in operational problems caused by deposits and slag during combustion.

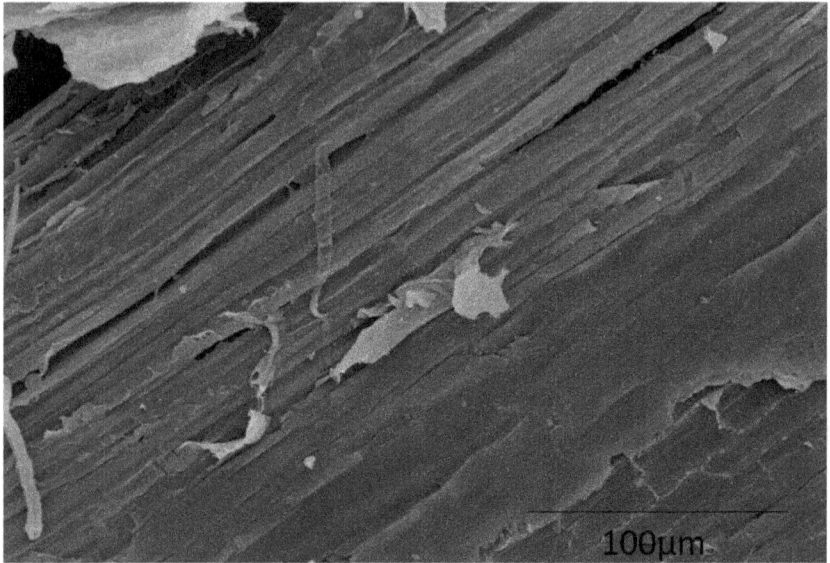

Figure 5.2 Micrograph of the inner surface of raw rice hull obtained by SEM.

Recently, a number of studies have examined the utilization of silica-based products as well as burning of rice husks to obtain energy and rice husk ash, an important silica source [41]. This material can be applied as a chromatography stationary phase [42], for producing nanoparticles [43] and composites [44], for contributing positive effects to the mechanical and durability performance when used as a supplementary cementitious material for building purposes, etc. [21,45-48].

Adsorptive properties of rice hull ash have been used for the removal of various compounds, such as dyes [49], heavy metals [50] and carbon dioxide [51]. Mor *et al.* [52] showed that activated rice husk ash could be effective for adsorbing phosphates from environments exposed to fertilizers and detergents. Rice hull ash has also been proposed to promote sustainable agriculture by protecting rice stems against pests [53]. Rice hull ash biochar can be applied to soils and offers numerous benefits [54]. The biochar can also be used for the removal of heavy metals from soil [55] and aqueous solutions [56].

Additionally, rice hull ash can be used as a biomaterial synthesis precursor (pseudowollastonite), which has biomedical properties. Silicon dioxide in rice hull ash has the potential to prevent bacterial colonization [57]. Rajanna *et al.* [58] also developed another bio-based material (silica aerogel microparticles) for drug delivery.

In the last several years, a number of literature studies have reported rice hulls as potential materials for cellulose and nanocellulose production [6,14,23,24,59-63]. According to Battegazzore *et al.* [6], rice hulls can be considered the most available cellulose source in the world due to its worldwide production. However, the high lignin and silica contents are important challenges. Lau *et al.* [63] reported that the use of pretreatments allows processes to overcome the rice hull lignin-hemicellulose barrier and to break down the silica shell.

5.3 Production of Nanocellulose from Rice Hulls

5.3.1 Cellulose Extraction

As a typical lignocellulosic residue, rice hulls are composed primarily of cellulose, hemicellulose and lignin. Cellulose and hemicellulose are structural polysaccharides, which typically make up two-thirds of the cell wall dry matter [64] with a cellulose:hemicellulose ratio that commonly varies between 2:1 to 1:1 [65]. Hemicellulose is grouped into four classes according to the main types of sugar residues present: xyloglucans, xylans, mannans and mixed-linkage β-glucans [66]. These groups of polysaccharides are insoluble in water, but can be solubilized in aqueous alkali media and hydrolyzed into component monosaccharides via diluted sulfuric acid (H_2SO_4). Hemicellulose is linked to cellulose and lignin via hydrogen and covalent bonding, respectively [67,68]. Lignin is generally the most complex and smallest fraction of the lignocellulosic biomass [68] and is a polyphenolic structural constituent of plants, representing the largest non-carbohydrate fraction of lignocellulose [64]. Lignin is considered an adhesive, holding cellulose and hemicellulose together [69], and is linked to both hemicellulose and cellulose to form a physical seal that is an impenetrable barrier in the plant cell wall, providing structural support, impermeability and resistance.

Cellulose is a semi-crystalline polymer that consists of β-D-glucopyranose units linked by β-1,4 glycosidic linkages forming linear long chains, which form elemental fibrils that are linked together by hydrogen bonds and van der Waals forces [70]. The cellulose molecules are organized as bundles of microfibrils consisting of highly ordered regions (crystalline fraction) alternating with less ordered regions (amorphous fraction). The less ordered regions are more accessible to attack by reagents and enzymes as well as absorption of water [67]. The glucose molecules and hydrogen bonding networks in cellulose have a wide orientation, resulting in four cellulose types (I, II, III and IV). Cellulose type I is the general allomorph of cellulose in nature, which constitutes the parallel packing of a hydrogen-bonded network [71-73]. The diameter of the elementary fibril is approximately 3.5 nm. The individual fibrils can align and bond together into nanofibrils (<35 nm), which can be bonded into fibrils (<1 μm) and fibers (10–50 nm) [72-73]. These cellulosic fibers are attached to each other by hemicellulose and covered by lignin [74].

Generally, the common steps employed to obtain nanocellulose from lignocellulosic residues are delignification and bleaching to obtain cellulose [60,75]. The purified cellulose is then subjected to an additional treatment to obtain nanocellulose, which varies based on the type of the desired nanocellulose.

Several processes have been reported as pretreatments to obtain cellulose from lignocellulosic materials. Generally, the structural arrangement of the lignocellulosic complex requires multiple step processes that include the use of pretreatments to disrupt the cellulose-hemicellulose-lignin complex to remove hemicellulose and lignin. This results in a cellulose-rich material that can be used after further purification for nanocellulose production [4,75-77].

The pretreatment methods can be divided as physical (milling, grinding and irradiation including gamma rays, electron beam, microwave irradiation, ultrasonication, etc.), physico-chemical (steam-explosion, hydrothermolysis and microwave-chemical pretreatment), biological (microorganisms and enzymes), chemical or a combination of these [67].

The most commonly employed chemical pretreatments of lignocellulosic residues are based on the use of acids or alkaline reagents. Acids can be used as catalysts for lignocellulosic material hydrolysis, as they can break down the heterocyclic ether bonds between the sugar monomers in the polysaccharides chains [78]. Thus, an acid pretreatment consists of the use of concentrated or diluted acids to break the rigid structure of the lignocellulosic material, solubilizing hemicellulose and releasing substantial amounts of sugars. The most commonly used acid is diluted H_2SO_4 [67].

The use of an alkaline pretreatment results in the partial depolymerization of lignin and hydrolysis of the hydrogen bonds between the -OH groups in hemicellulose, thus, representing this approach as an effective method to remove lignin and hemicellulose. The pretreatment also partially removes the impurities present on the surface of the fibers, such as pectins, waxes, and oils [79]. Wang *et al.* [80] reported that sodium hydroxide (NaOH) can be used in alkaline pretreatments with concentrations ranging from 1 to 10% (g/g dry matter). Saponification and cleavage of lignin-carbohydrate linkages is the main function of the alkaline pretreatment, which increases the porosity and internal surface area of the cellulosic fibers as well as decreases the degree of polymerization and crystallinity. Consequently, all these processes favor the penetration of the bleaching agents into the fiber's structure [81].

In recent years, our research group has been working on the extraction of cellulose from several lignocellulosic residues. In a previous study [24], we obtained cellulose from rice hulls with a goal of nanocellulose production using an alkaline pretreatment (NaOH 5% w/v) followed by peracetic acid bleaching for cellulose purification. However, the bleaching of rice hulls was shown to be a very challenging process compared to other lignocellulosic residues, and many physical, chemical and combination processes have been tested. Some preliminary results are presented in this chapter.

Figure 5.3 shows the effect of an alkaline pretreatment with NaOH (5% w/v) for 60 min at 90 °C on the morphology of the rice hull surface. Compared with Figure 5.1, it can be observed that the outer surface composed mainly of hemicellulose, lignin and silica was removed, resulting in cellulosic fibers that were linked together and were accessible to other chemical, physical or enzymatic agents. The surface composition of this sample, determined by EDS, exhibited a silica content of 1.1%, indicating that the alkaline pretreatment with NaOH was effective in removing silica from rice hulls. Lau *et al.* [63] reported that silica and lignin act as external armor and block other components from reacting with the external environment, and also that the hydroxide solutions are effective in disrupting the silica layer from rice hulls.

Peracids are another group of chemical reagents being studied for delignification and cellulose pulp bleaching and have shown good results in terms of resistance, indicating low levels of cellulose degradation. We note that this is a completely chlorine-free technique, which is an advantage as compared to conventional bleaching methods that employ chlorites and hypochlorites and result in toxic chlorinated effluents. Peracetic acid can be prepared by the oxidation of acetic acid by hydrogen peroxide. It is considered as a highly selective

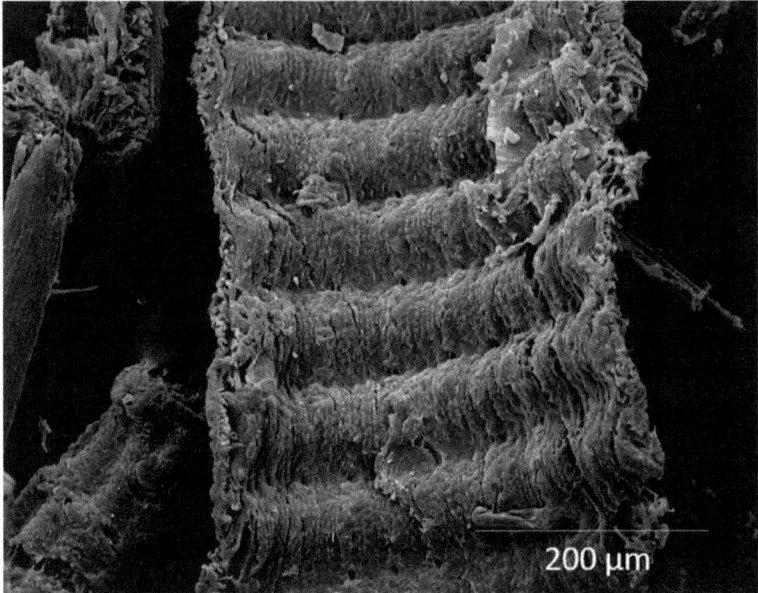

Figure 5.3 Micrograph obtained by SEM of rice hull after subjecting it to an alkaline pretreatment with 5% NaOH.

delignification agent due to its capacity to oxidize structures rich in electrons, such as the aromatic rings of lignin. During the treatment of a pulp with peracetic acid, opening of the aromatic ring makes the oxidized lignin more hydrophilic, contributing to its solubilization in the bleaching liquor. In addition, the formation of acid groups also favors the solubilization of fragments of lignin, especially during the alkaline extraction stages [82]. Paschoal et al. [83] reported that the use of peracetic acid to obtain cellulose from oats was effective in removing lignin, resulting in individualized bundles of cellulose microfibrils when examined with SEM.

Figure 5.4 shows a SEM micrograph of a rice hull after bleaching with peracetic acid at 60 °C for 24 h. It can be observed that the surface did not change significantly, and silica remained on the outer surface. Also, the external layer was observed to be broken at some points. However, the use of peracetic acid was ineffective in removing lignin and silica from rice hulls, necessitating an additional treatment to remove the remaining silica and lignin. The surface composition of the sample, determined by EDS, resulted in a silica content of 20%, which was very close to the content in raw rice hulls. As reported above and according to Nascimento et al. [24], the combination of an alkaline pretreatment followed by bleaching with peracetic acid was effective for obtaining cellulose from rice hulls.

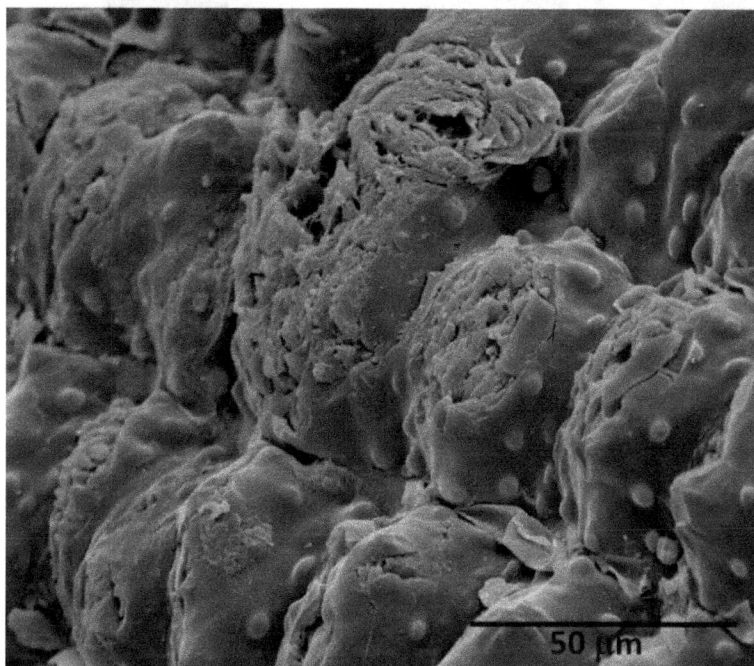

Figure 5.4 SEM micrograph of rice hull after bleaching with peracetic acid at 60 °C for 24 h.

Figure 5.5 presents the micrograph of a rice hull after bleaching with peracetic acid at 60 °C for 24 h twice. After second treatment with peracetic acid, rice hull presented a smoother surface without silica, with some pores and cracks (arrows), however, this treatment was ineffective in removing lignin and hemicellluloses. The surface composition of the sample, determined by EDS, resulted in a silica content of 4%.

Figure 5.6 presents the scanning electron micrograph of rice hull after subjecting to an alkaline pretreatment with NaOH (5% w/v) over 60 min at 90 °C followed by a bleaching step with peracetic acid at 60 °C for 24 h. The combination of the treatments was effective in removing silica, lignin and hemicelluloses from rice hulls, resulting in individualized cellulosic fibers, which can be used for isolation of nanocellulose. The surface composition of this sample by EDS analysis exhibited a silica content of lower than 1 %.

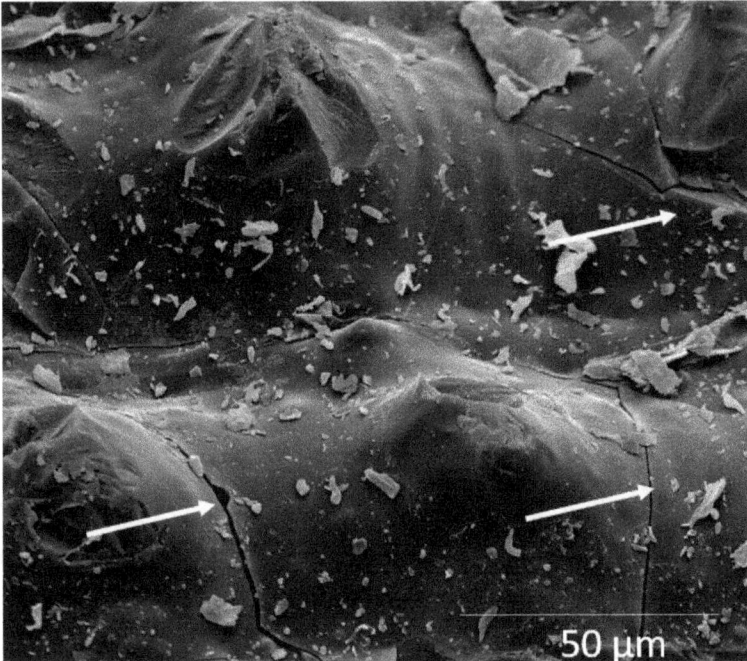

Figure 5.5 Scanning electron micrograph of rice hull after bleaching with peracetic acid at 60 °C for 24 h twice.

Battegazzore *et al.* [6] reported the extraction of cellulose from rice hulls through a three consecutive step process. The first step was an acid pretreatment with H_2SO_4 to hydrolyze the hemicellulose and to remove impurities. After filtration, the solid residue was subjected to an alkaline treatment with potassium hydroxide (KOH) for silica extraction. After filtration, bleaching with sodium chlorite ($NaClO_2$) was performed to remove the amorphous cellulose and lignin.

Oliveira *et al.* [61] obtained cellulose from rice hulls by employing an alkaline pretreatment with NaOH (4% w/v, 80 °C, 4 h) for seven times for lignin and hemicellulose removal. Subsequently, the solid residue was subjected to a bleaching step with a mixture of NaOH, acetic acid and $NaClO_2$ at 4 h at 90 °C. The purified cellulose contained 93.1% cellulose, 4.2% hemicellulose, 0.8% lignin and 1.9% ash.

5.3.2 Nanocellulose Isolation

Nanocellulose is defined as the cellulose material smaller than 100 nm in at least one dimension [84-87]. Nanocellulose isolated from plant sources can be classified into cellulose nanocrystals or nano-fibrillated cellulose according to their morphology. Bacterial cellulose is also classified as nanocellulose.

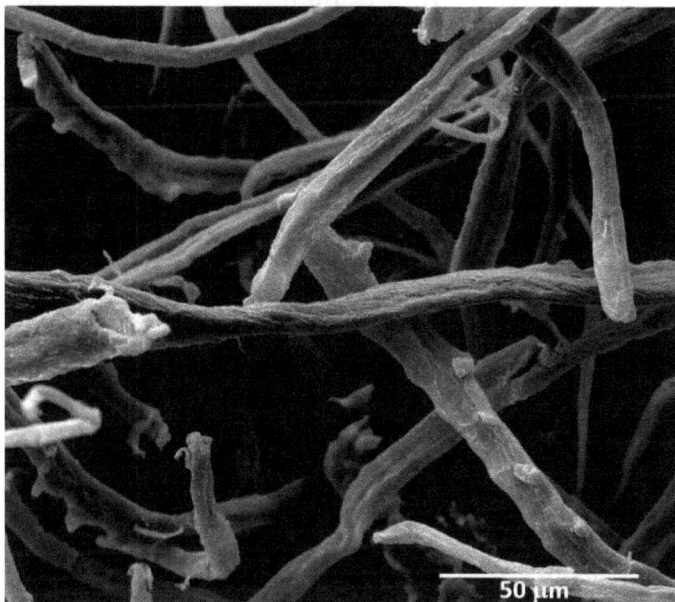

Figure 5.6 Scanning electron micrograph of rice hull after subjecting to an alkaline pretreatment with 5% NaOH followed by a bleaching step with peracetic acid at 60 °C for 24 h.

Cellulose nanocrystals or whiskers have a perfect crystalline structure with a crystallinity index of 54% to 82%. They have short-rod-like or whisker shapes with diameters ranging from 3 to 35 nm and lengths of 200 to 500 nm [71,73]. Nanofibrillated cellulose or cellulose nanofibers are materials consisting of flexible and entangled cellulose fibers with diameters ranging from 10 to 100 nm and lengths of several micrometers. These have both crystalline and amorphous alternating fractions [72].

Cellulose nanocrystals were first isolated in 1949 via acid hydrolysis with H_2SO_4 [88]. For this, the amorphous fraction of cellulose is hydrolyzed and removed by acid while maintaining the crystalline fraction [73]. Until today, the typical protocol for the isolation of cellulose nanocrystals has consisted of several stages after acid hydrolysis with H_2SO_4 (generally employed at a concentration of 64%), including repeated washing steps with centrifugation and dialysis with distilled water to remove any remaining free acid molecules. Then, a mechanical treatment, such as sonication, is required to disperse the nanocrystals in a uniform and stable suspension [87]. The process is complex and requires many days. Additionally, a large amount of effluents results from the washing steps, and the yields can vary widely.

Nanofibrillated cellulose production by mechanical action under heat was first reported in 1983 [89]. Several mechanical processes to obtain nanofibrillated cellulose have been reported in the literature based on the cleavage of cellulose fibrils along the longitudinal axis. Conventional processes, such as high-pressure homogenization, grinding and microfluidization, and non-conventional processes, such as extrusion, steam explosion, ball milling, cryocrushing and high-intensity ultrasonication, have been reported for nanocellulose production. These processes are efficient in disintegrating the cell wall, but require intensive mechanical treatments and high-energy consumption due to the cohesion of the cell wall. Using conventional mechanical treatments without any pretreatments results in inhomogeneous materials that may contain poorly fibrillated fibers. In addition, this process is associated with high costs. This limitation can be overcome by combining several sequential chemical and mechanical treatments [85,90-92].

Ludueña *et al.* [23] produced nanocellulose with an average diameter of 12.4 nm and a crystallinity index of 76%. In this process, rice hulls were subjected to sequential chemical pretreatments including the use of an alkaline medium with KOH (3% w/v) followed by an acid medium with hydrochloric acid (HCl acid 10% v/v) for removal of silica. However, lignin and waxes still remained in the material. Therefore, the material was treated with sodium chlorite ($NaClO_2$, 0.70% w/v) and sodium bisulfite ($NaHSO_3$, 5% w/v) followed by an alkaline treatment with NaOH (17% w/v). Finally, nanocellulose was obtained by acid hydrolysis with H_2SO_4 (60% w/v).

Johar *et al.* [14] reported the production of cellulose nanocrystals from rice hulls after subjecting the material to an alkaline pretreatment with NaOH (4% w/v) thrice to remove lignin and hemicellulose, followed by bleaching by adding a buffer solution of acetic acid, aqueous chlorite (1.7% w/v) and distilled water at reflux. This process was repeated four times, resulting in an increase in the cellulose content from 31% (raw rice hull) to 96%. Cellulose nanocrystals were successfully extracted from the purified cellulose from rice hulls using an acid hydrolysis treatment with H_2SO_4 (10 mol/L, 50 °C, 40 min), resulting in the material with diameter ranging from 10 to 15 nm and length less than 100 nm, along with a crystallinity index of 59%.

Rosa *et al.* [59] reported the use of a chlorine-free procedure to isolate cellulose from rice hulls, resulting in non-toxic effluent and a cellulose yield of 74%, which exhibited a crystallinity index of 67%. The delignification was performed at 121 °C in an autoclave using NaOH (5% w/v, for 30 min), followed by an ultrasonic treatment for 30 min. To remove the remaining hemicellulose and lignin, the resulting pulp was bleached using a two-step bleaching process with hydrogen peroxide/tetraacetylethylenediamine (TAED). Subsequently, to purify the cellulose pulp, acetic acid (80% v/v) and nitric acid (70%, v/v) were used to produce a total yield of of 28 wt%. The cellulosic fibers were subjected to acid hydrolysis with H_2SO_4 (64% w/w, 25 °C, 60 min), resulting in nanocrystals in the form of needles ranging from 100 to 400 nm in length and 6 to 14 nm in diameter.

Nascimento *et al.* [24] isolated nanofibrillated cellulose from rice hulls via bleaching with NaOH (5% w/v, 90 °C, 60 min), followed by bleaching with a peracetic acid solution (50% acetic acid, 38% hydrogen peroxide 120 V and 12% distilled water) at 60 °C for 24 h. After bleaching, the compact structure around the cellulosic fibers was removed, and the lignin content of the residue decreased from 7.22 to 4.22%. The cellulosic fibers were subjected to acid hydrolysis with H_2SO_4 (64% w/w, 45 °C, 60 and 120 min), followed by ultrasonication for 15 min. The obtained nanofibrillated cellulose exhibited higher crystallinity (70%) and superior thermal stability than the raw material, along with a lignin content below 0.35%. The nanocellulose formed interconnected webs of tiny fibers, with diameters less than 100 nm and lengths of several micrometers.

According to Islam *et al.* [60], the large-scale production of cellulose nanocrystals from lignocellulosic materials, such as rice hulls, has not been widely explored or established, and at a laboratory scale, the cellulose yields are also very low (<30%). The authors reported the use of a 2.5 L reactor to perform a low pressure alkaline delignification of the rice hulls, followed by a bleaching process with a sodium hypochlorite solution (NaClO), resulting in a material with 95% cellulose and 52.2% crystallinity index. The cellulose nanocrystals were obtained by acid hydrolysis with H_2SO_4 (4 mol/L, 60 °C, 60 min), resulting in particles with diameters of 15-50 nm and lengths of 275-550 nm.

Collazo-Bigliardi *et al.* [62] employed an alkaline pretreatment with NaOH (4% w/v, 3 h) twice, followed by bleaching with $NaClO_2$

(1.7% w/v, 4 h) for four times, to obtain a cellulose-rich sample. The composition of the resulting sample was 73.8% cellulose, 19% hemicellulose, 1.6% lignin and 0.14% ash, along with a total yield of 41%. This cellulose-rich sample was employed to obtain cellulose nanocrystals by acid hydrolysis with H_2SO_4 (64% w/w, 50 °C, 40 min), resulting in nanocrystals with a rod-like structure, lengths of 310±160 nm, diameters of 39±13 nm and a crystallinity index of 90%.

5.4 Final Considerations

The use of rice hulls for the production of cellulose and nanocellulose is aligned with the worldwide trend of developing a sustainable economy based on raw materials from renewable sources. Considering the high world production of rice, which is distributed over many different countries and regions, it becomes even more attractive to use this raw material for the extraction of cellulose, as it is one of the most available cellulose sources. There are some challenges involved in the use of lignocellulosic materials for obtaining cellulose and nanocellulose. Especially, the structural arrangement of the cellulose-hemicellulose-lignin complex and variations among these three components depending on the plant species and other climatic and management conditions make it difficult to generalize pretreatments and protocols for cellulose and nanocellulose production.

In the last several years, many research studies have reported the use of rice hulls to obtain cellulose and nanocellulose. However, this material presents additional challenges compared to other lignocellulosic materials due to its higher silica content. Some multi-step protocols have been described in the literature, which require the use of several chemicals reagents as well as neutralization, washing and purification steps, along with the generation of a significant amount of effluents.

It is clear that this topic still requires further research and development to obtain satisfactory results by overcoming the obstacles, especially by identifying methods that produce less pollution, which will permit large-scale production of both cellulose and nanocellulose.

References

1. Bhowmick, G. D., Sarmah, A. K., and Sen, R. (2018) Lignocellulosic

biorefinery as a model for sustainable development of biofuels and value added products. *Bioresource Technology*, **247**, 1144-1154.

2. Zuin, V. G., and Ramin, L. Z. (2018) Green and sustainable separation of natural products from agro-industrial waste: challenges, potentialities, and perspectives on emerging approaches. *Topics in Current Chemistry*, **3**, 1-54.

3. Islam, M. S., Kao, N., Bhattacharya, S. N., Gupta, R., and Choi, H. J. (2018) Potential aspect of rice husk biomass in Australia for nanocrystalline cellulose production. *Chinese Journal of Chemical Engineering*, **26**, 465-476.

4. Gupta, A., and Verma, J. P. (2015) Sustainable bio-ethanol production from agro-residues: a review. *Renewable and Sustainable Energy Reviews*, **41**, 550-567.

5. Valdebenito, F., Pereira, M., Ciudad, G., Azocar, L., Briones, R., and Chinga-Carrasco, G. (2017) On the nanofibrillation of corn husks and oat hulls fibres. *Industrial Crops and Products*, **95**, 528-534.

6. Battegazzore, D., Bocchini, S., Alongi, J., Frache, A., and Marino, F. (2014) Cellulose extracted from rice husk as filler for poly (lactic acid): preparation and characterization. *Cellulose*, **21**, 1813-1821.

7. *Circular Series – Word Agricultural Production*, United States Department of Agriculture (2018). Online: https://apps.fas.usda.gov/psdonline/circulars/production.pdf [accessed 10th January 2019].

8. *FAOSTAT Data*, Food and Agriculture Organization of the United Nations (2016). Online: http://www.fao.org/faostat/en/#home [accessed 13th July 2018].

9. *Save and Grow in Practice: Maize, Rice and Wheat, a Guide to Sustainable Cereal Production*, Food and Agriculture Organization of the United Nations (2016). Online: http://www.fao.org/3/a-i4009e.pdf [accessed 14th July 2018].

10. Maraseni, T. N., Deo, R. C., Qu, J., Gentle, P., and Neupane, P. R. (2018) An international comparison of rice consumption behaviours and greenhouse gas emissions from rice production. *Journal of Cleaner Production*, **172**, 2288-2300.

11. Syuhadah, S., and Rohasliney, H. (2012) Rice husk as biosorbent: A review. *Health and the Environment Journal*, **3** (1), 89-95.

12. Quispe, I., Navia, R., and Kahhat, R. (2017) Energy potential from rice husk through direct combustion and fast pyrolysis: A review. *Waste Management*, **59**, 200-210.

13. Bhammathidas, N., and Mehta, P. K. (2004) Concrete Mixtures Made with Ternary Blended Cements Containing Fly Ash and Rice Husk Ash. *CANMET*, India, pp. 379-391.

14. Johar, N., Ahmad, I., and Dufresne, A. (2012) Extraction, preparation and characterization of cellulose fibres and nanocrystals from rice husk. *Industrial Crops and Products*, **37**, 93-99.

15. Yoshida, S. (1981) *Fundamentals of Rice Crop Science*. IRRI, Philip-

pines.
16. Chandrasekhar, S., Satyanarayana, K. G., Pramada, P. N., Raghavan, P., and Gupta, T. N. (2003) Processing, properties and applications of reactive silica from rice husk – An overview. *Journal of Materials Science*, **38**, 3159-3186.
17. Jackson, M. G. (1977) Review article: the alkali treatment of straws. *Animal Feed Science and Technology*, **2**, 105-130.
18. Jacometti, G. A.; Mello, L. R. P. F., Nascimento, P. H. A., Sueiro, A. C., Yamashita, F., and Mali, S. (2015) The physicochemical properties of fibrous residues from the agro Industry. *LWT - Food Science and Technology*, **62**, 138-143.
19. Potumarthi, R., Baadhe, R. R., Nayak, P., and Jetty, A. (2013) Simultaneous pretreatment and sacchariffication of rice husk by *Phanerochete chrysosporium* for improved production of reducing sugars. *Bioresource Technology*, **28**, 113-117.
20. Shahabazuddin, Md., Sarat Chandra, T., Meena, S., Sukumaran, R. K., Shetty, N. P., and Mudliar, S. N. (2018) Thermal assisted alkaline pretreatment of rice husk for enhanced biomass deconstruction and enzymatic saccharification: Physico-chemical and structural characterization. *Bioresource Technology*, **263**, 199-206.
21. Thomas, B. S. (2018) Green concrete partially comprised of rice husk ash as a supplementary cementitious material – a comprehensive review. *Renewable and Sustainable Energy Reviews*, **82**, 3913-3923.
22. Pérez, J., Muñoz-Dorado, J., De-la-Rubia, T., and Martínez, J. (2002) Biodegradation and biological treatments of cellulose, hemicellulose and lignin: An overview. *International Microbiology*, **5**(2), 53-63.
23. Ludueña, L., Fasce, D., Alvarez, V. A., and Stefani, P. M. (2011) Nanocellulose from rice husk following alkaline treatment to remove silica. *BioResources*, **6**(2), 1440-1453.
24. Nascimento, P. H. A., Marim. R., Carvalho, G. M., and Mali, S. (2016) Nanocellulose produced from rice hulls and its effect on the properties of biodegradable starch films. *Materials Research*, **9**(1), 167-174.
25. Akhtar, M., Bhanger, M. I., Iqbal, S. and Hasany, S.M. (2006) Sorption potential of rice husk for the removal of 2,4- dichlorophenol from aqueous solutions: kinetic and thermodynamic investigations. *Journal of Hazardous Materials*, **128**(1), 44-52.
26. Naiya, T. K., Bhattacharya, A. K., and Das, S. K. (2009) Adsorptive removal of Cd(II) ions from aqueous solutions by rice husk ash. *Environmental Progress and Sustainable Energy*, **28**(4), 535-546.
27. Naiya, T. K., Bhattacharya, A. K., Mandal, S. and Das, S. K. (2009) The sorption of lead(II) ions on rice husk ash. *Journal of Hazardous Materials*, **163**(2-3), 1254-1264.

28. Chuah, T. G., Jumasiah, A., Azni, I., Katayon, S., and Thomas-Choong, S. Y. (2005) Rice husk as a potentially low-cost biosorbent for heavy metal and dye removal: an overview. *Desalination*, **175**(3), 305-316.
29. Guiso, M. G., Alberti, G., Emma, G., and Pesavento, M. (2012) Pb (II), Cu (II) and Cd (II) removal through untreated rice husk; thermodynamics and kinetics. *Analytical Sciences*, **28**, 993-999.
30. Ye, H., Zhu, Q. and Du, D. (2010) Adsorptive removal of Cd (II) from aqueous solution using natural and modified rice husk. *Bioresource Technology*, **101**(14), 5175-5179.
31. Li, W. C., Law, F. Y., and Chan, Y. H. M. (2017) Biosorption studies on copper (II) and cadmium (II) using pretreated rice straw and rice husk. *Environmental Science and Pollution Research*, **24**, 8903-8915.
32. Saha, B. C., and Cotta, M. A. (2008) Lime pretreatment, enzymatic saccharification and fermentation of rice hulls to ethanol. *Biomass and Bioenergy*, **32**, 971-977.
33. Dagnino, E. P., Chamorro, E. R., Romano, S. D., Felissia, F. E., and Area, M. C. (2013) Optimization of the acid pretreatment of rice hulls to obtain fermentable sugars for bioethanol production. *Industrial Crops and Products*, **42**, 363-368.
34. Moscon, J. M., Priamo, W. L., Bilibio, D., Silva, J. R. F., Souza, M., Lunelli, F., Foletto, E. L., Kuhn. R. C., Cancelier, A., and Mazutti, M. A. (2014) Comparison of conventional and alternative technologies for the enzymatic hydrolysis of rice hulls to obtain fermentable sugars. *Biocatalysis and Agricultural Biotechnology*, **3**, 149-154.
35. Gupta, R., Pandit, A., Nirjar, A., and Gupta, P. (2013) Husk power systems: bringing light to rural India and tapping fortune at the bottom of the pyramid. *Asian Journal of Management Cases*, **10**, 129-143.
36. Mustonen, S., Raiko, R., and Luukkanen, J. (2013) Bioenergy consumption and biogas potential in cambodian households. *Sustainability*, **5**, 1875-1892.
37. Shackley, S., Carter, S., Knowles, T., Middelink, E., Haefele, S., and Haszeldine, S. (2012) Sustainable gasification-biochar systems? A case-study of rice-husk gasification in Cambodia, part II: Field trial results, carbon abatement, economic assessment and conclusions. *Energy Policy*, **41**, 618-623.
38. Yoon, S. J., Son, Y. -I., Kim, Y. -K, and Lee, J. -G. (2012) Gasification and power generation characteristics of rice husk and rice husk pellet using a downdraft fixed-bed gasifier. *Renewable Energy*, **42**,163-167.
39. Abu Bakar, M. S., and Titiloye, J. O. (2013) Catalytic pyrolysis of rice husk for bio-oil production. *Journal of Analytical and Applied Pyrolysis*, **103**, 362-368.
40. Bazargan, A., Bazargan, M., and McKay, G. (2015) Optimization of rice husk pretreatment for energy production. *Renewable Energy*,

77, 512-520.

41. Bakar, R. A., Yahya, R., and Gan, S. N. (2016) Production of high purity amorphous silica from rice husk. *Procedia Chemistry*, **19**, 189-195.

42. Shahnani, M., Mohebbi, M., Mehdi, A., Ghassempour, A., and Aboul-Enein, H. Y. (2018). Silica microspheres from rice husk: A good opportunity for chromatography stationary phase. *Industrial Crops and Products*, **121**, 236-240.

43. Sankar, S., Kaur, N., Lee, S., and Kim, D. Y. (2018) Rapid sonochemical synthesis of spherical silica nanoparticles derived from brown rice husk. *Ceramics International*, **44**(7), 8720-8724.

44. Real, C., Córdoba, J. M., and Alcalá, M. D. (2018) Synthesis and characterization of SiC/Si_3N_4 composites from rice husks. *Ceramics International*, **44**(12), 14645-14651.

45. Zareei, S. A., Ameri, F., Dorosktar, F., and Ahmadi, M. (2017) Rice husk ash as a partial replacement of cement in high strength concrete containing micro silica: evaluating durability and mechanical properties. *Case Studies in Construction Materials*, **7**, 73-81

46. Adenuga, O. A., Soyingbe, A. A., Ogunsanmi, O. E. (2010) The use of rice husk ash as partial replacement for cement in concrete. *Lagos Journal of Environmental Studies*, **7**(2), 47-50.

47. Antiohos, S. K., Papadakis, V. G., and Tsimas, S. (2014) Rice husk ash (RHA) effectiveness in cement and concrete as a function of reactive silica and fineness. *Cement and Concrete Research*, **61-62**, 20-27.

48. Andreola, F., Lancellotti, I., Manfredini, T., Bondioli, F., and Barbieri, L. (2018) Rice husk ash (RHA) recycling in brick manufacture: effects on physical and microstructural properties. *Waste and Biomass Valorization*, **9**(12), 2529-2539.

49. Barbosa, T. R., Foletto, E. L., Dotto, G. L., and Jahn, S. L. (2018) Preparation of mesoporous geopolymer using metakaolin and rice husk ash as synthesis precursors and its use as potential adsorbent to remove organic dye from aqueous solutions. *Ceramics International*, **44**(1), 416-423.

50. Maingi, F. M., Mbuvi, H. M., and Mwangi, H. (2018) Adsorption of cadmium ions on geopolymers derived from ordinary clay and rice husk ash. *International Journal of Materials and Chemistry*, **8**(1), 1-9.

51. Gargiulo, N., Shibata, K., Peluso, A., Aprea, P., Valente, T., Pezzotti, G., Shiono, T., and Caputo, D. (2018) Reinventing rice husk ash: derived NaX zeolite as a high-performing CO_2 adsorbent. *International Journal of Environmental Science and Technology*, **15**(7), 1543-1550.

52. Mor, S., Chhoden, K., and Ravindra, K. (2016) Application of agro-waste rice husk ash for the removal of phosphate from the wastewater. *Journal of Cleaner Production*, **129**, 673-680.

53. Jeer, M., Suman, K., Maheswari, T. U., Voleti, S. R., and Padmakumari,

A. P. (2018) Rice husk ash and imidazole application enhances silicon availability to rice plants and reduces yellow stem borer damage. *Field Crops Research,* **224**, 60-66.

54. Liu, S., Meng, J., Jiang, L., Yang, X., Lan, Y., Cheng, X., and Chen, W. (2017) Rice husk biochar impacts soil phosphorous availability, phosphatase activities and bacterial community characteristics in three different soil types. *Applied Soil Ecology*, **116**, 12-22.

55. O'Connor, D., Peng, T., Li, G., Wang, S., Duan, L., Mulder, J., Cornelisse, G., Cheng, Z., Yang, S., and Hou. D. (2018) Sulfur-modified rice husk biochar: A green method for the remediation of mercury contaminated soil. *Science of The Total Environment,* **621**, 819-826.

56. Xiang, J., Lin, Q., Cheng, S., Guo, J., Yao, X., Liu, Q., Guangcai, Y., and Liu, D. (2018) Enhanced adsorption of Cd (II) from aqueous solution by a magnesium oxide–rice husk biochar composite. *Environmental Science and Pollution Research*, **25**(14), 14032-14042.

57. Azam, F. A. A., Shamsudin, R., Ng, M. H., Ahmad, A., Akbar, M. A. M., and Rashidbenam, Z. (2018) Silver-doped pseudowollastonite synthesized from rice husk ash: Antimicrobial evaluation, bioactivity and cytotoxic effects on human mesenchymal stem cells. *Ceramics International*, **44**(10), 11381-11389.

58. Rajanna, S. K., Kumar, D., Vinjamur, M., and Mukhopadhyay, M. (2015) Silica aerogel microparticles from rice husk ash for drug delivery. *Industrial & Engineering Chemistry Research*, **54**(3), 949-956.

59. Rosa, S. M. L., Rehman, N., Miranda, M. I. G., Nachtigall, S. M. B., and Bica, C.I.D. (2012) Chlorine-free extraction of cellulose from rice husk and whisker isolation. *Carbohydrate Polymers*, **87**, 1131-1138.

60. Islam, M. S., Kao, N., Bhattacharya, S. N., Gupta, R., and Bhattacharjee, P. K. (2017) Effect of low pressure alkaline delignification process on the production of nanocrystalline cellulose from rice husk. *Journal of the Taiwan Institute of Chemical Engineers*, **80**, 820-834.

61. Oliveira, J. O., Bruni, G. P., Lima, K. O., El Halal, S. L. M., Rosa, G. S., Dias, A. R. G., and Zavareze, E. R. (2017) Cellulose fibers extracted from rice and oat husks and their application in hydrogel. *Food Chemistry*, **160**, 153-160.

62. Collazo-Bigliardi, S., Ortega-Toro, R. and Boix, A. C. (2018) Isolation and characterisation of microcrystalline cellulose and cellulose nanocrystals from coffee husk and comparative study with rice husk. *Carbohydrate Polymers*, **191**, 205-215.

63. Lau, B. B. Y, Yeung, T., Patterson, R. J., and Aldo, L. (2017) A cation study on rice husk biomass pretreatment with aqueous hydroxides: cellulose solubility does not correlate with improved enzymatic hydrolysis. *ACS Sustainable Chemistry & Engineering*, **5**, 5320-5329.

64. Mounika, M., and Ravindra, K. (2015) Characterization of nanocomposites reinforced with cellulose whiskers: A review. *Materials Today: Proceedings*, **2**, 3610-3618.

65. Schadel, C., Blöchl, A., Richter, A., and Hoch, G. (2010) Quantification and monosaccharide composition of hemicelluloses from different plant functional types. *Plant Physiology and Biochemistry*, **48**(1), 1-8.

66. Alila, S., Besbes, I., Vilar, M. R., Mutjé, P., and Boufi, S. (2013) Non-woody plants as raw materials for production of microfibrillated cellulose (MFC): A comparative study. *Industrial Crops and Products*, **41**(1), 250-259.

67. Menon, V., and Rao, M. (2012) Trends in bioconversion of lignocellulose: biofuels, platform chemicals biorefinery concept. *Progress in Energy and Combustion Science*, **38**(4), 522-550.

68. Anwar, Z., Gulfraz, M., and Irshad, M. (2014) Agro-industrial lignocellulosic biomass a key to unlock the future bio-energy: a brief review. *Journal of Radiation Research and Applied Sciences*, **7**(2), 163-173.

69. Ummartyotin, S., and Manuspiya, H. (2015) A Critical review on cellulose: from fundamental to an approach on sensor technology. *Renewable and Sustainable Energy Reviews*, **41**, 402-412.

70. Nechyporchuk, O., Belgacem, M. N., and Bras, J. (2016) Production of cellulose nanofibrils: a review of recent advances. *Industrial Crops and Products*, **93**, 2-25.

71. Moon, R. J., Martini, A., Nairn, J., Simonsen, J., and Youngblood, J. (2011) Cellulose nanomaterials review: structure, properties and nanocomposites. *Chemical Society Reviews*, **40**, 3941-3994.

72. Ling, S., Chen, W., Fan, Y., Zheng, K., and Jin, K. (2018) Biopolymer nanofibrils: Structure, modeling, preparation, and applications. *Progress in Polymer Science*, **85**, 1-56.

73. Phanthong, P., Reubroycharoen, P., Hao, X., Xu, G., Abudula, A., and Guam, G. (2018) Nanocellulose: Extraction and application. *Carbon Resources Conversion*, **1**, 32-43.

74. Kalia, S., Boufi, S., Celli, A., and Kango, S. (2014) Nanofibrillated cellulose: surface modification and potential applications. *Colloid and Polymer Science*, **292**(1), 5-31.

75. Abdul-Khalil, H. P. S, Davoudpour, Y., Saurabh, C. K., Hossain, M. S., Adnan, A. S., Dungani, R., Paridah, M. T., Islam Sarker, M .Z., Nurul Fazita, M. R., Syakir, M. I., and Haafiz, M. K. M. (2016) A review on nanocellulosic fibres as new material for sustainable packaging: process and applications. *Renewable and Sustainable Energy Reviews*, **64**, 823-836.

76. Zhang, X., Tu, M., and Paice, M. G. (2011). Routes to potential bioproducts from lignocellulosic biomass lignin and hemicelluloses. *BioEnergy Research*, **4**(4), 246-257.

77. Garcia-Maraver, A., Salvachúa, D., Martínez, M. J., Diaz, L. F., and Zamorano, M. (2013) Analysis of the relation between the cellulose, hemicellulose and lignin content and the thermal behavior of resi-

dual biomass from olive trees. *Waste Management*, **33**(11), 2245-2249.

78. Laopaiboon, P., Thani, A., Leelavatcharamas, V., and Laopaiboon, L. (2010) Acid hydrolysis of sugarcane bagasse for lactic acid production. *Bioresource Technology*, **101**, 1036-1043.
79. Ng, H. M., Sin, L. T., Tee, T. T., Bee, S. T., Hui, D., Low, C. Y., and Rahmat, A. R. (2015) Extraction of cellulose nanocrystals from plant sources for application as reinforcing agent in polymers. *Composites, Part B: Engineering*, **75**, 176-200.
80. Wang, S., Li, F., Zhang, P., Jin, S., Tao, X., Tang, X., Ye, J., Nabi, M., and Wang, H. (2017) Ultrasound assisted alkaline pretreatment to enhance enzymatic saccharification of grass clipping. *Energy Conversion and Management*, **149**, 409-415.
81. Kallel, F., Bettaieb, F., Khiari, R, García, A., Bras, J., and Chaabouni, S. E. (2016) Isolation and structural characterization of cellulose nanocrystals extracted from garlic straw residues. *Industrial Crops and Products*, **87**, 287-296.
82. Brasileiro, L. B., Colodette, J. L., and Piló-Veloso. (2001) A utilização de perácidos na deslignificação e no branqueamento de polpas celulósicas. *Quimica Nova*, **24** (6), 819-829.
83. Paschoal, G., Muller, C. M. O., Carvalho, G. M., Tischer, C. A., and Mali, S. (2015) Isolation and characterization of nanofibrillated cellulose from oat hulls. *Quimica Nova*, **38**(4), 478-482.
84. Isogai, A., and Bergström, L. (2018) Preparation of cellulose nanofibers using green and sustainable chemistry. *Current Opinion in Green and Sustainable Chemistry*, **12**, 15-21.
85. Lindstrom, T. (2017) Aspects on nanofibrillated cellulose (NFC) processing, rheology and NFC-film properties. *Current Opinion in Colloid and Interface Science*, **29**, 68-75.
86. Qu, X., Alvarez, P. J. J., and Li, Q. (2013) Applications of nanotechnology in water and waste water treatment. *Water Research*, **47**, 3931-3946.
87. Brinchi, L., Cotana, F., Fortunati, E., and Kenny, J. M. (2013) Production of nanocrystalline cellulose from lignocellulosic biomass: technology and applications. *Carbohydrate Polymers*, **94**(1), 154-169.
88. Ranby, B. G. (1949) Aqueous colloidal solutions of cellulose micelles. *Acta Chemica* Scandinavica, **3**, 649-650.
89. Turbak, A. F., Snyder F. W., and Sandberg, K. R. (1983) Microfibrillated Cellulose, A New Cellulose Product: Properties, Uses, and Commercial Potential. *Ninth Cellulose Conference*, USA, pp. 815-827.
90. Naderi, A., Koschella, A., Heinze, T., Shih, K. C., Nieh, M. P., Pfeifer, A., and Chang, C. C., and Erlandsson, J. (2017) Sulfoethylated nanofibrillated cellulose: production and properties. *Carbohydrate Polymers*, **169**, 515-523.

91. Boufi, S., and Chaker, A. (2016) Easy Production of cellulose nanofibrils from corn stalk by a conventional high speed blender. *Industrial Crops and Products*, **93**, 39-47.

92. Chauhan, V. S., and Chakrabarti, S. K. (2012) Use of nanotechnology for high performance cellulosic and papermaking products. *Cellulose Chemistry* and *Technology*, **46**(5-6), 389-400.

6

Nanocellulose: An Advanced Green Material

Saleheen Bano,* Asif Ali, Sauraj and Yuvraj Singh Negi

Department of Polymer and Process Engineering, IIT Roorkee Saharanpur Campus, Saharanpur 247001, India

Corresponding author: s.bano.24.2009@gmail.com

6.1 Introduction

Innovation, development and utilization of different composite materials for various applications are continuously attracting the researchers and scientists to improve the life style of mankind. Among these, green and sustainable composite materials are becoming more fascinating in the world of materials science, as they can circumvent the issues related to environment and sustainability to a major extent.

In the recent years, development and advancement of nanomaterials have attracted growing interest due to a variety of potential applications. However, the production of synthetic nanomaterials occasionally requires the use of expensive and harmful chemicals obtained from petroleum-based sources, which adversely affect the environment [1]. Thus, high performance renewable products of biological origin have been gaining attention as source of nanomaterials owing to their bio-degradability, relative abundance, cost effectiveness, biocompatibility, ease of processability and excellent mechanical properties [2,3].

In this context, cellulose, the major component of plant cell wall and the most abundant natural polymer, finds its potential for the development of nanocellulose, which exhibits promising applications in various industrial sectors. The nanocellulose (NC)-based materials, having at least one dimension in nanometer range, possess surpassing features of high aspect ratio, large surface area and superior mechanical properties along with biocompatibility, non-toxicity, chemical resistivity and availability. Also, the presence of

Nanocellulose, edited by Vikas Mittal
© 2019 Central West Publishing, Australia

surface hydroxyl group provides hydrophilicity and ease for surface modification. The hydrogen bonding interaction of NCs with the polymer matrices along with high stiffness provide attractive properties to the nanocomposite materials. Cellulose nanomaterials have the potential to meet almost all the requirements for being known as green nanomaterial [3-5].

Nanocellulose produced either by mechano-chemical, enzymatic or acid hydrolysis methods is mainly classified as cellulose nanofibers (CNFs) and cellulose nanocrystals (CNCs) also known as cellulose whiskers (CWs), which exhibit differences in the aspect ratio, morphology and crystallinity [6]. Both types of NCs are frequently utilized in various applications, the discussion in present chapter is mainly focused on nano-crystalline celluloses.

The cheapest sources of cellulosic material are the waste or by-products from agricultural crops that are produced globally every year at a large scale, and a major portion of these wastes is burnt as fuel, resulting into environmental issues. Thus, there is a growing interest for transformation of these materials into value added products (i.e. CNCs), which results in significant contribution towards waste management [7].

A large number of studies have been carried out on the production of different types of NCs from different sources, along with applications. The present chapter sheds light on different aspects of NCs, particularly nano-crystalline celluloses, which make them sustainable materials of choice for 21st century.

6.2 Overview of Cellulose

6.2.1 Background

Cellulose is the most abundant natural polymer on earth, and the annual production of cellulose was estimated as 1.5×10^{12} tons [8]. It is mainly present in the secondary cell wall of plants along with lignin, hemicelluloses, pectin, waxes and other extractives. The compositional value of the components varies depending on the origin, species and the environment in which the plant grows [6]. The lingo-cellulosic biomass is considered as the rich source of cellulose, however, some bacteria, algae, fungi and marine animals like tunicates also produce cellulose in small amount [5].

Anselme Payen carried out the extraction of cellulose fiber from the plant tissues for the first time in 1838 [9]. Subsequently, cellulose and its derivatives attracted the attention of researchers and industries from various disciplines. In 1868, Hyatt Manufacturing Company led the production of cellulose from lingo-cellulosic sources at industrial scale [8]. Its polymeric nature was first determined by Staudinger in 1920, after which it gained the interest of researchers in the field of polymer chemistry [10]. First chemical synthesis of cellulose was performed by Kobayashi and Shoda in 1992 without using any biologically derived enzymes [11].

6.2.2 Structural Aspect and Properties

Cellulose is an odorless, semi-crystalline, biodegradable, non-toxic, hydrophilic and chemically inert biomaterial with high strength and modulus. The physical and chemical aspects of cellulose are largely governed by its supramolecular structure [12], which is described briefly in this section.

In the plant cell wall, cellulose is not found as an isolated molecule, but is present as assembly of fibers. These cellulosic fibers, in turn, are composed of several micro-fibrils which are an assembly of elementary nano-fibrils consisting of cellulosic chains with both crystalline and amorphous regions [5,13]. The strong inter- and intramolecular hydrogen bonding present between the cellulosic chains generates crystallinity in the fibers [3]. However, the reason for the presence of amorphous region is attributed to chain dislocation, as explained by some theoretical models reported previously [14].

Although the structure of cellulose considerably depends on the sources of origin, however, regardless of source, it is commonly characterized as high molecular weight, linear and homo-polysaccharide with cellobiose as repeat unit, which is a dimer of anhydro-β, D-glucopyranoses linked together with β-1, 4-glycosidic linkages (Figure 6.1). The cellulose chain at its one end consists of non-reducing hemi-acetal unit, whereas it possesses anomeric carbon with free hydroxyl group at the other end for the formation of glycosidic bonds. This configuration produces chemical asymmetry in the cellulose chain. As a general consideration, the cellulose chains consist of up to 20,000 repeat units, though cellulose chains with lesser number of repeat units are also present, which are mainly embedded in the primary cell walls [8,15, 8,16].

As mentioned previously [3,17], the three hydroxyl groups positioned equatorially at C-2, C-3 and C-6 carbons are easily available for hydrogen bonding between the cellulosic chains. The complex network of hydrogen bonds not only results in the formation of highly crystalline domains (where the cellulose chains are tightly packed in ordered fashion), but also contributes high mechanical strength and insolubility to cellulose fibers. These crystalline domains can be further isolated as cellulose nanocrystals with excellent properties.

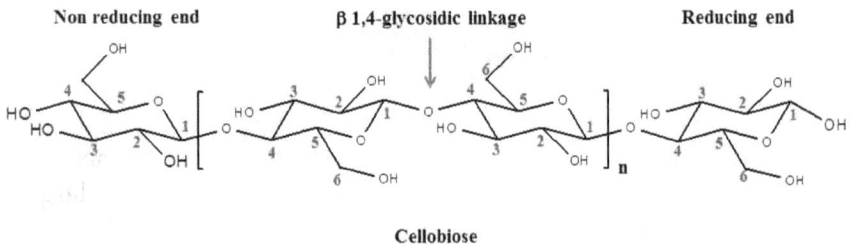

Figure 6.1 Chemical structure of cellulose macromolecule.

Based on molecular orientation, location of hydrogen bonds, extraction and treatment methods, cellulose exists in different types of inter-convertible polymorphs or allomorphs, such as cellulose I, II, III and IV [6,15,18]. The most common crystalline and thermodynamically metastable form in which the native cellulose exists is cellulose I. It has been further revealed that two distinct phases namely cellulose I_α and cellulose I_β coexist together in cellulose I with varying proportion. The I_α form is prevalent in bacterial and algal cellulose, while I_β is present in large proportion in the higher plants [19].

Nishiyama *et al.* [20] revealed that both cellulose I_α and I_β comprised of parallel configuration in their structure, but differed in their hydrogen bond patterns, which resulted in the variation in the crystalline structure (I_α and I_β correspond to triclinic and monoclinic unit cells respectively). The former phase consists of one chain per unit cell, whereas the latter has two chains per unit cell.

In cellulose I_α, the parallel chains are piled together via van der Waals interactions with progressive shear oriented parallel to the chain axis, whereas cellulose I_β has stack of parallel cellulose chains with alternating shear. Cellulose I_α is thermodynamically metastable and can be converted into more stable I_β form without losing its crystallinity, by hydrothermal annealing or by treatment with different solvents [21].

Moreover, Kim *et al.* [22] revealed the inter-conversion of different cellulose forms. For instant, cellulose I can be converted into cellulose II by alkali treatment, which exists as monoclinic phase with two chains per unit cell and has anti-parallel configuration defined by packing of chains with opposite polarity (i.e. chains are packed with their reducing ends on opposite sides within crystalline domains) as compared to cellulose I which comprises of parallel chain arrangement (having their reducing ends on the same side). The cellulose III form is generated by the treatment of cellulose I or II with liquid ammonia or various ammines, which leads to the formation of cellulose III$_I$ or III$_{II}$ depending on the starting form. The cellulose III$_I$ possesses one chain per monoclinic unit cell symmetry with parallel chain configuration, as cellulose I. However, the exact structure of cellulose III$_{II}$ is not yet established. The study by Wada *et al.* [23] suggested it to be disordered phase of cellulose with one chain unit cell comprising of anti-parallel configuration. Another polymorph defined by Gardiner and Sarco [24] is cellulose IV, which is considered as a result of heat treatment of cellulose II or III at 260 °C in glycerol. It cannot be transformed directly from cellulose I.

Overall, cellulose I is the most common form of cellulose fibers with high crystallinity and axial elastic modulus (130 GPa to 250 GPa), which provides mechanical reinforcement to the polymer matrices. As mentioned earlier, the stiffness of cellulose fibers may be explained by intra- and intermolecular hydrogen bonding network between cellulose chains, as shown in Figure 6.2. It has been reported that intramolecular hydrogen bonds contribute high modulus to cellulose as compared to intermolecular bonds, and between the two intramolecular hydrogen bonds, i.e. $O(6)...H-O(2)'$ and $O(3)-H...O(5)'$, the hydrogen bond between hydroxyl group at C-3 position and pyranose ring oxygen atom $O(5)'$ of other unit in the same chain is responsible for the fiber stiffness [25,26].

The chemical inertness and insolubility of cellulose in various solvents depend on its degree of crystallinity. The crystalline part of cellulose micro-fibrils contains tightly packed orderly arrangement of cellulose chains which does not allow the chemical reagents to pass through and reach the reactive sites. On the other hand, the amorphous part (with loosely packed disordered cellulose chains) is less durable and prone to the chemical attack. However, the accessibility of cellulose can be enhanced by chemical modification. The incorporation of charges leads to the separation of chains which increase its solubility and, thus, applicability in diverse fields [12].

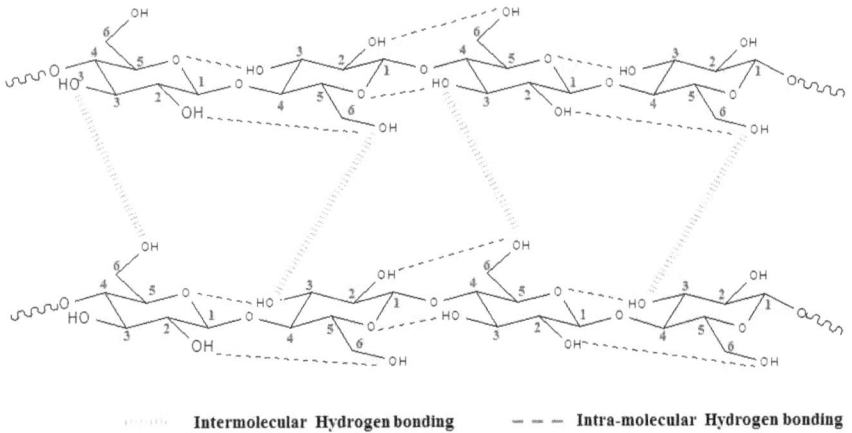

······· Intermolecular Hydrogen bonding   – – – Intra-molecular Hydrogen bonding

Figure 6.2 Hydrogen bond network present in the cellulose molecule.

The isolation of value added nanocellulose-based materials requires extraction of high purity cellulose fibers from the lingo-
cellulosic mass through different treatments (described in later sections). It, thus, emphasizes the need of maximizing the yield of pure
cellulose by removing other constituents from the hierarchal structure of lingo-cellulosic fibers [27].

6.3 Classification of Nanocellulose

As mentioned earlier, nanostructured celluloses have received significant research attention owing to their attractive features. The
versatility and potential of these nanostructures in different applications has been reported in previous studies [28].

Nanocellulose is a generic term used for the isolated cellulose
with at least one dimension in the nanometric range. As also mentioned earlier, nanocelluloses can be classified into three categories
based on the source, isolation technique, morphology and properties, viz. CNCs, CNFs and bacterial nanocellulose (BNC) (Figure 6.3)
[9]. The first two are isolated from agricultural sources (woody and
non-woody) through top-down process by chemical, mechanical and
enzymatic treatments, whereas the last one is produced by microbes
through bottom-up process.

As per the new TAPPI standard (WI 3021), CNCs are defined as
pure crystalline nanostructure having dimensions of 3-10 nm in
width and aspect ratio ranging from 5-50, while CNFs have dimen-

sions of 5-30 nm in width and aspect ratio of more than 50 with both crystalline and amorphous regions in the molecular chains. The BNCs are basically cellulose nano-fibrils produced with high purity and crystallinity through the conversion of glucose molecules by different species of bacteria such as acetobacter, agrobacterium, gluconobacter and rhizobium [29]. BNCs present interlaced ribbon like structure having diameter of 100 nm and length of few micrometers.

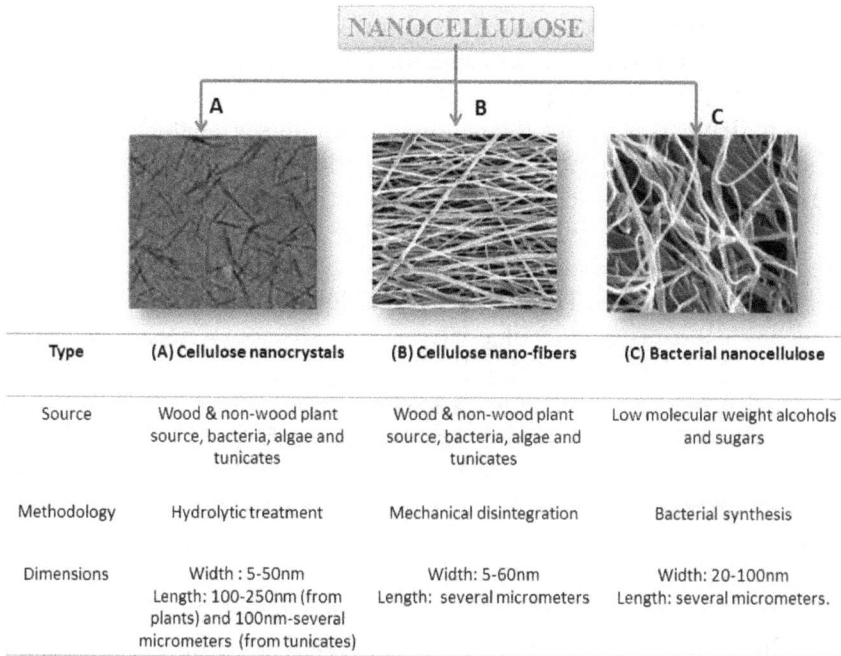

Type	(A) Cellulose nanocrystals	(B) Cellulose nano-fibers	(C) Bacterial nanocellulose
Source	Wood & non-wood plant source, bacteria, algae and tunicates	Wood & non-wood plant source, bacteria, algae and tunicates	Low molecular weight alcohols and sugars
Methodology	Hydrolytic treatment	Mechanical disintegration	Bacterial synthesis
Dimensions	Width : 5-50nm Length: 100-250nm (from plants) and 100nm-several micrometers (from tunicates)	Width: 5-60nm Length: several micrometers	Width: 20-100nm Length: several micrometers.

Figure 6.3 Classification of nanocellulose.

The production of CNCs involves hydrolytic degradation method, especially acid hydrolysis of cellulose fibers, which removes the amorphous part, leaving the crystalline part intact. In case of CNFs, different types of mechanical treatments, such as homogenization, ultrasonication and grinding with discs, help to disintegrate cellulose microfibers into nanofibers. In few instances, chemical or biological treatment of cellulose fibers prior to mechanical disintegration has been carried out to reduce the energy cost for the production of CNFs [30,31]. BNCs are obtained by fermentation of culture medium rich in mono- and oligosaccharides and further isolated by alkali treatment. The highly pure BNCs, without any traces of lignin, hemicelluloses and pectin, are the most viable material for medical applications [32].

6.4 Cellulose Nanocrystals

The cellulose micro-fibrils comprise of crystalline domains linked by amorphous regions, which upon processing through hydrolysis breakdown into highly crystalline nanoparticles. Acid hydrolysis is the most commonly employed method for the isolation of cellulose nanocrystals, however, enzymes and some ionic liquids have also been reported for the generation of CNCs. The exact mechanism for acid hydrolysis is yet to be established, though it is considered that the amorphous region is sensitive to acid attack and allows the protonation of glycosidic oxygen as well as successive dissociation of glycosidic linkages, which results in the dissolution of amorphous regions leaving behind the rod like highly crystalline nano-domains [33,34] (Figure 6.4). These nano-crystalline celluloses have superior mechanical, optical and electrical properties over native cellulose fibers, which emphasize the need to produce these at larger scales to fabricate nanocomposites for various applications.

In 1950, Ranby, for the first time, prepared a stable colloidal suspension of cellulose, obtained from wood, by using sulfuric acid. The author further reported the first transmission electron microscopy (TEM) images of the dried suspensions revealing the presence of needle like cellulose particles which possessed same crystalline nature as parent cellulose when analyzed by electron diffraction [35,36]. Today, such particles are well known as CNCs. The liquid crystalline property of the aqueous suspension of CNCs was manifested in 1959, however, the mechanism of the formation of stable chiral nematic liquid crystalline phase was later reported in 1992 by Revol *et al.* [37]. Since the first study by Favier *et al.* [38] reporting the significant improvement in the mechanical properties of CNC reinforced composites, extensive research effort has been focused on CNC based nanocomposites for various applications.

6.4.1 Isolation of Cellulose Nanocrystals

Among the various sources of cellulose, the agricultural wastes find their place as most attractive and economic source. For example, sugarcane bagasse [39], rice husk [40], pineapple leaves [41], grape

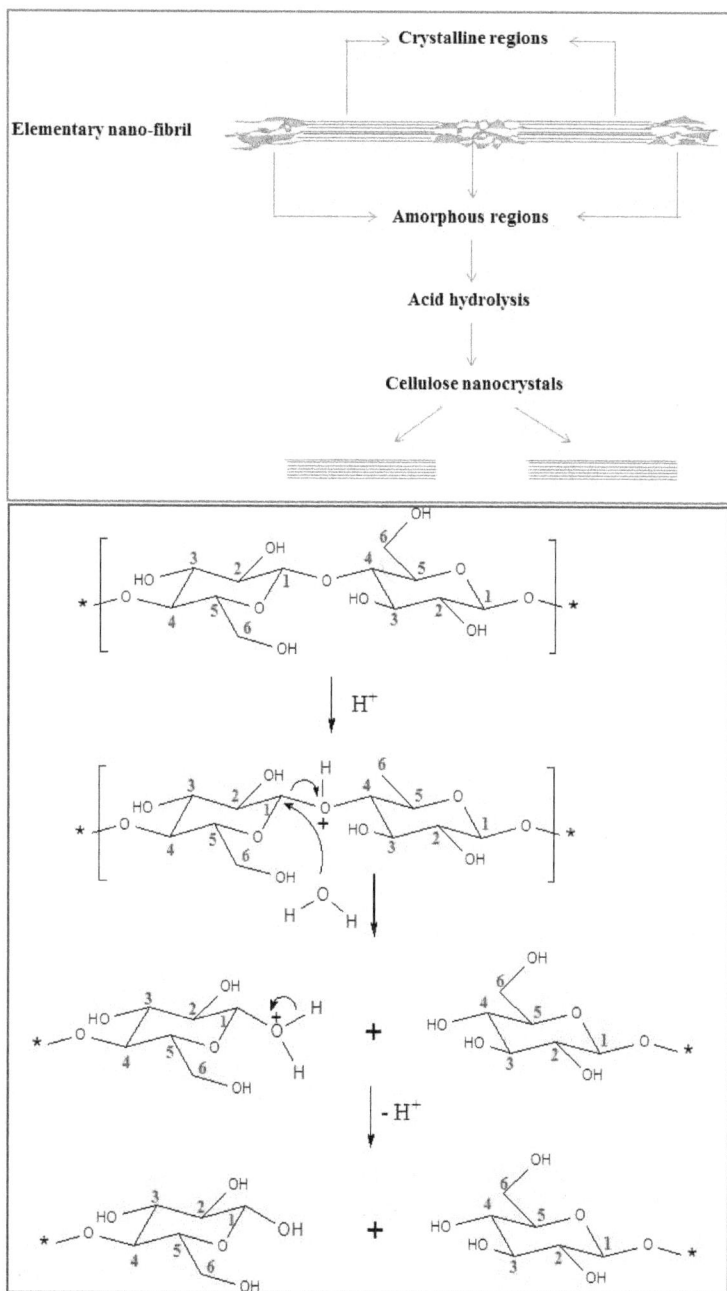

Figure 6.4 Schematic representation of (A) acid hydrolysis of elementary cellulose nanofiber and (B) proposed mechanism for the hydrolysis process.

skin [42], groundnut shells [7], garlic straw [43], etc., have been uti-lized for cellulose and nanocellulose extraction. These collectively account for significant extent of agricultural wastes, which are ei-ther burnt or used as cattle feed. For the last decades, the frequent use of these waste materials for the isolation of nanocellulose has helped to enhance their market value. Figure 6.5 schematically pre-sents the general procedure of CNC isolation from biomass.

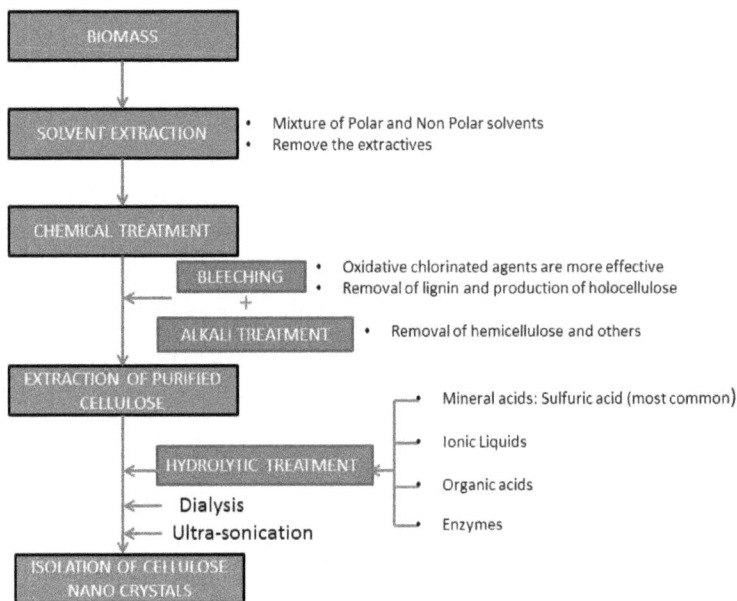

Figure 6.5 Schematic representation of CNC extraction from biomass.

Solvent Extraction

Ligno-cellulosic biomass contains various types of extractives, such as fatty acids, waxes, terpenoids, flavonoids, rosins, tannins, etc., along with structural components, i.e. cellulose, hemicellulose and lignin [44]. In order to obtain pure cellulose, the removal of these extractives from biomass is essential prior to chemical treatments. Solvent extraction technique is preferably used for this process, which requires combination of different polar and non-polar sol-vents like methanol-benzene [7,45], ethanol-toluene [46], etc., to dissolve the extractives. The extraction is generally carried out for 8-10 h.

Chemical Treatments: Isolation of Purified Cellulose

During bleaching or delignification process, lignin and some hemicellulose get removed. The complex formed between lignin and other carbohydrates resists the process of hydrolysis, thus, there is a need to remove lignin from the complex. Various oxidative and reductive bleaching agents are used during the process, however, mainly chlorinated oxidative agents exhibit better efficiency for the bleaching process [44].

The use of eco-friendly methods for the fractionation of lingocellulosic materials into their constitutive components has gained significant research attention as the use of chlorinated products during delignification or bleaching process generates hazardous by-products. In this regard, various non-chlorinated products, like alkaline hydrogen peroxides, have been studied [47]. Table 6.1 summarizes the different chlorinated and non-chlorinated bleaching agents used in the process.

For isolation of pure cellulose, alkali treatment with 1-2 M sodium hydroxide (NaOH) or potassium hydroxide (KOH) has been observed to dissolve hemicellulose, pectin and other impurities.

Table 6.1 Types of oxidative bleaching agents for delignification process

Bleaching Agents	
Chlorinated compounds	**Non-chlorinated compounds**
Sodium hypochlorite	Hydrogen peroxide
Sodium chlorite	Sodium per borate
Calcium hypochlorite	Sodium per carbonate
Chlorine dioxide	Sodium hydrosulfite
	Sodium bisulfite

Acid Hydrolysis: Extraction of Cellulose Nanocrystals

The isolated purified cellulose through chemical treatments is further subjected to hydrolytic process for extraction of CNCs. The resulting hydrolyzed cellulose suspension is subsequently diluted with excess amount of water and subjected to dialysis, followed by centrifugation and sonication to obtain suspension of pure CNCs.

As mentioned earlier, acid hydrolysis has been commonly studied to extract CNCs, which facilitates the cleavage of accessible glycosidic linkages within the amorphous region of cellulose microfibrils, resulting in rod shaped nanocrystals (Figure 6.4). Concentrated sulfuric acid is widely employed for CNCs extraction, as it imparts anionic sulfate groups on crystal surfaces that develop electrostatic repulsion between nano-crystalline particles, yielding stable CNC suspensions in water.

In addition to source materials, different hydrolysis conditions including temperature, hydrolysis duration, acid concentration and acid to pulp ratio have significant impact on the properties of CNCs, namely dimensions and crystallinity degree [48,49]. However, it has been concluded that the temperature has a major influence on the surface charge compared to other factors. Though CNCs obtained after sulfuric acid hydrolysis exhibit high stability in aqueous suspensions, but these suffer from lower stability due to presence of sulfate groups [50], which restricts the use of CNCs in some bio-nanocomposites.

Besides sulfuric acid, other mineral acids such as hydrochloric acid, phosphoric acid, hydro-bromic acid and their mixtures have also been studied for the hydrolysis process [15,51,52]. Despite high efficiency for the isolation of CNCs, acid assisted hydrolysis has some limitations due to the hazardous, corrosive and bio-incompatible nature of strong mineral acids. To overcome these, alternative processes like enzymatic hydrolysis, ionic liquid mediated hydrolysis, weak organic acids-based hydrolysis, etc., have also been proposed in recent years. An example of the extraction of cellulose and CNCs from groundnut shells is shown in Figure 6.6.

6.5 Properties of Cellulose Nanocrystals

6.5.1 Inherent Properties

The renewable and biocompatible nano-crystalline celluloses with high surface area (250-500 m^2/g), good thermal stability (onset degradation ranges from 200°C-300 °C) [53] and crystallinity (in range of 60-90%) [27,54] have been mostly reported to possess rod shaped morphology, however CNCs with spherical morphology have also been reported [55]. The dimensions of rod or needle like CNCs vary from 100 to 1000 nm in length and from 5 to 30 nm in width, depending on the origin and hydrolysis conditions.

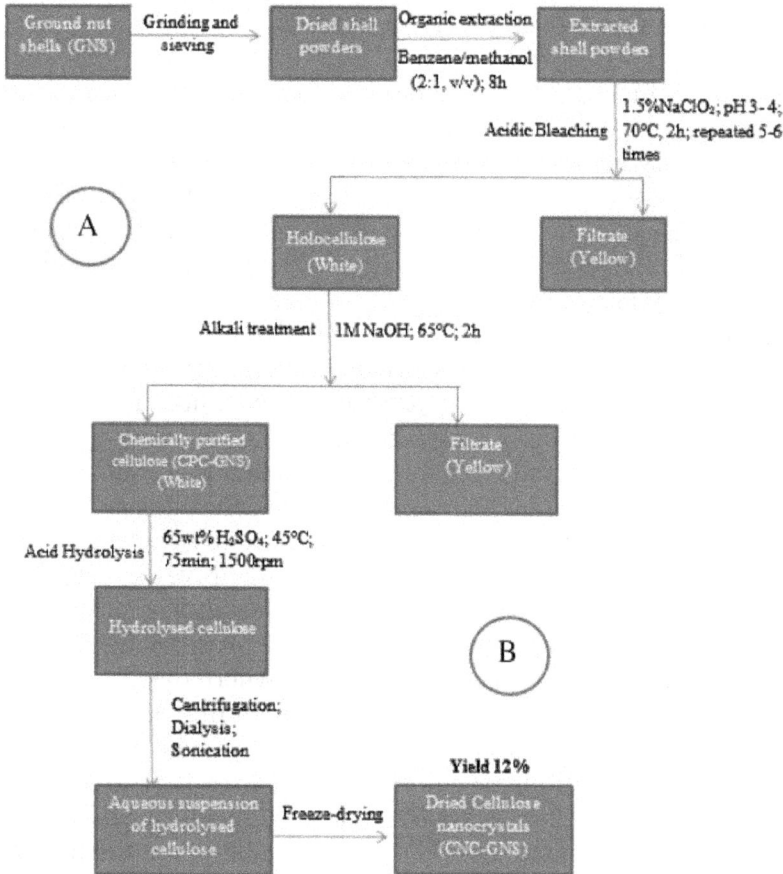

Figure 6.6 Schematic presentation for isolation of cellulose and CNCs from ground nut shells. Reproduced from Reference 7 with permission from Elsevier.

CNCs with high aspect ratio are essential for achieving improved mechanical properties. It has been reported that nanocellulose materials with aspect ratio 10 or more are considered as potential reinforcing fillers [56,57]. The elastic modulus (100-140 GPa) and strength (7.5 GPa) of CNCs (obtained from various sources), measured using different techniques including Raman spectroscopy, X-ray diffraction and atomic force microscopy, were reported to be high enough as comparable Kevlar and steel [3]. Apart from these properties, suspensions of CNCs in different media and in the form of thin solid films possess liquid crystalline chiral nematic properties.

6.5.2 Liquid Crystalline Behavior of CNCs

Apart from their inherent properties, CNCs in liquid suspensions (generated using different media) and in solid thin films (obtained after slow drying of CNC suspensions) were reported to have chiral nematic phase which provides them uniquely colored appearance without incorporation of pigments. This peculiar feature results in their use as templates for the development of photonic materials in decorative and security papers.

Heux *et al.* [58] reported for the first time nematic crystalline structure with pitch value of 4 micron for surfactant coated CNCs in non-polar solvents, which was attributed to enhanced dispersability and steric stabilization exerted by the surfactant coating on CNC surface. In another study, the authors reported that non-polar suspensions of CNCs having high aspect ratio produced anisotropic gel phase instead of showing any phase separation [59].

In general, the polarized optical microscopy reveals that the cellulose nanocrystals are randomly oriented at dilute concentrations in aqueous suspensions, forming an isotropic phase. However, as the concentration increases, a stage is reached where the CNCs get self-organized in a particular direction resulting in the formation of liquid crystalline anisotropic phase, identified as chiral nematic or chlolestic liquid crystal structures by Revol *et al.* [60]. These structures consist of stacked layers of suspension in which CNC rods are aligned in a preferential direction (director), however, the direction of alignment of each plane is rotated with respect to previous plane, thus, giving rise to helical structure, where pitch is defined as vertical distance required to complete 360° rotation of the director.

Chiral nematic order of aqueous suspensions of CNCs get enhanced by the application of magnetic field, as cellulose is diamagnetic in nature and the nanorods tend to align perpendicular to the magnetic field [61]. Kvien and Oksman [62] attempted to align CNCs in a polymer matrix (poly(vinyl alcohol) (PVA)) by using a strong magnetic field to obtain a unidirectional reinforced nanocomposite. Interestingly, the results showed that the dynamic modulus of the nanocomposite was higher in the aligned direction compared with the transverse direction. Colloidal suspensions of CNCs from ramie fibers in cyclohexane have also been oriented in an AC electric field [63] as revealed by TEM and electron diffraction. Moreover, these suspensions demonstrated a birefringence under polarized light, whose magnitude could be increased with increasing field strength.

Under controlled conditions, the slow evaporation of water leads to the formation of thin solid films in which the chiral nematic liquid crystalline order of suspension is preserved. These films present iridescence by reflecting the polarized light in a specific range of wavelength. The wavelength of reflected light and, hence, the perceived color are dependent on the helical pitch, angle of incident light and refractive index of the material [48].

The liquid crystallinity of CNCs or the chiral nematic pitch depends on various factors such as concentration, aspect ratio and dispersability of CNCs, electrolyte strength, external stimuli such as magnetic field and temperature, nature and density of charges present on the CNCs (which is dependent on the type of acid used for hydrolysis) etc. [64]. Interesting optical properties can be achieved for CNCs-based nanocomposites, which can be tuned further with the described factors.

6.5.3 Optical Properties

In addition to mechanical reinforcement, the optical transparency of CNCs-based nanocomposites is another important aspect which is determined by the transmittance of light in the visible region. It has been revealed that higher the transmittance, better will be the transparency of composites, which is dependent on the size and dispersability of CNCs in the polymer matrices. CNCs with high aspect ratio form a continuous network which has negative impact on the transparency of the polymer matrix, since it promotes the scattering of light to a major extent [65].

6.5.4 Barrier Properties

Cellulose nanocrystals exhibit superior barrier properties owing to their nano-dimensions, high crystallinity and ability to form a network which provide a tortuous path to the permeates (gases, water vapor, liquid, etc.) and reduce their diffusivity through polymer matrices in the nanocomposites. However, the gas permeability and water diffusion coefficient of neat cellulose nanocrystal films were reported to be higher than that of nano-fibrillated cellulose (NFC) films, thus, exhibiting inferior barrier properties. It could be ascribed to an increase in the tortuosity of diffusion path due to large fibril entanglement and lower porosity of NFC films. Moreover, presence of residual lignin and other extractives on NFC surface can possibly reduce the swelling of these films in water. The presence of sulfate groups on the CNCs surface creates osmotic pressure, which separates the crystals and allows water penetration, thus, resulting in higher degree of swelling [66,67].

In case of poly(lactic acid) (PLA) based systems, a decrease in water transmission and oxygen permeability was reported by Sanchez-Garcia and Lagaron [68] with addition of up to 5 wt% of CNCs, which was attributed to the nanocrystal induced crystallization of PLA. The CNCs reinforced nanocomposites of hydrophilic polymers such PVA, chitosan and carboxy methyl cellulose (CMC) were observed to show lower water vapor transmission rate (WVTR) than neat polymer due to the tortuosity effect of CNCs [69].

A hybrid nanocomposite of natural rubber (NR) reinforced with CNCs and montmorillonite (MMT) was reported by Bendahou *et al.* [70]. The composite exhibited improved gas barrier properties owing to synergistic effect of CNCs and MMT nanoparticles. It was reported that with MMT concentration higher than 2.5 wt%, the barrier efficiency increased effectively. The CNCs have also been used for the preparation of ultrafine membranes for water purification due to permselectivity towards positively charge ions. The permselectivity can be ascribed to the presence of negatively charged sulfate groups on CNCs surface which repel and do not allow the transport of other negatively charged species through the membrane [71].

6.5.5 Rheological Properties

Cellulose nanocrystals possess unique rheological features. In recent years, investigations have been carried out on the rheological properties of CNCs and their nanocomposites to correlate microstructural changes and related outcomes during processing, handling and application [72].

Rheology of CNC Suspensions

It has been revealed that two phase transitions appear in the rheological behavior of a CNCs suspension: from isotropic to lyotropic liquid crystal (LC) and from liquid crystal to gel as the concentration of CNCs increases in the aqueous suspension. The flow behavior of a CNCs suspension gets significantly affected by the physico-chemical properties of the particles. The appearance of phase transition in a CNCs suspension is largely governed by the concentration, dimensions and surface charge density of CNCs, temperature and ionic strength of the system, etc. [48,73-75].

It has been observed that the viscosity of a CNCs suspension depends on the concentration, shear rate and aspect ratio of CNCs. Viscosity increases with an increase in the concentration due to enhanced possibility of CNCs collisions [76]. However, at dilute concentrations (up to 3 wt%), the aqueous suspension of CNCs (isotropic phase) exhibits shear thinning behavior, where the viscosity decreases with increase in shear rate. Above a critical concentration, CNCs suspension exhibits liquid crystalline phase where the flow properties of the CNCs suspension comprises of three well defined regions in the viscosity profile, similar to the anomalous behavior of the LC polymer solution [73]. These three regions can be assigned as (1) region I: at low shear rates (< 1 s^{-1}), the suspension shows effective shear thinning characteristic where LCs tend to align under flow, (2) region II: at moderate shear rates (1-10 s^{-1}), the viscosity decrease with shear rate is less noticeable and can be attributed to more orderly arrangement of LCs in the shear direction due to particle interactions (these interactions develop resistance to the flow and result in the shear rate independent plateau region) and (3) region III: at high shear rates (>10 s^{-1}), the orderly arranged liquid crystalline structure of CNCs suspension gets destroyed, and the crystalline domains become freed, defining a second shear thinning behavior by essential alignment of crystals in the flow direction. With further increase in concentration, another critical concentration is obtained above which the suspension behaves like a rheological gel showing single shear thinning behavior at all shear rates [77]. According to previous studies [73,75], the temperature also renders significant influence on the rheological characteristics owing to the variation in relative percentage of isotropic and lyotropic liquid crystalline phases.

Wu *et al.* [78] concluded that the suspension of CNCs with higher aspect ratio undergoes biphasic transition and forms hydrogel at lower concentration as compared to the suspension with low aspect ratio CNCs. Moreover, CNCs with higher aspect ratio render high relaxation time constant and, thus, remain aligned for longer time period even after removal of shear forces [17].

In addition, the rheological properties of CNCs suspension are dependent on the type of acid used for hydrolysis. The suspension of CNCs obtained after sulfuric acid treatment shows occasional independent shear thinning behavior, whereas the HCl treated CNCs

suspension displays thixotropy and anti-thixotropy at correspondingly higher and lower concentrations, along with higher shear
thinning behavior [79,80].

Rheological Characteristics of CNCs-based Polymer Nanocomposites

The CNCs embedded between the polymeric chains can help to mod-
ify the rheology of resulting nanocomposites. Studies on rheological
properties of polymer nanocomposites are crucial for understand-
ing their processing characteristics by gaining information about the
(1) dispersion of nanofiller, (2) interaction between nanofiller and
polymer, (3) filler-filler interactions and (4) interfaces in polymer
nanocomposites (structure-property relationships).

Several literature studies have described the influence of various
aspects of CNCs (such as concentration, morphology, surface area
and surface chemistry), temperature and polymer matrix on the
rheological properties of resulting nanocomposites. It has been reported that the polymer-CNCs nanocomposites exhibit higher viscosity than the neat matrix, and it increases with increase in CNCs
content. The viscoelastic behavior is evaluated by measuring storage modulus (G'), loss modulus (G") and tan delta. Studies reveal that
G' and G" increase with increase in the CNCs content but the increase
in G' is more than G" [81,82]. At lower concentrations (~1 wt%),
loss modulus is higher as compared to storage modulus and the
nanocomposites behave like viscoelastic fluid, while at higher concentrations (>1 wt%), the storage modulus possess higher values
than loss modulus and the nanocomposites display gel like behavior.

The transition from viscous fluid to gel like solid material can also be demonstrated by examining dissipation factor (tan delta). A
value <1, at higher CNCs content, implies dominant elasticity in the
material, whereas samples with lower CNCs content behave as viscous fluid with tanδ value >1. The complex viscosity (η^*) of the system is also observed to enhance with increase in CNCs content [83].

Similar observations were reported for the systems having CNCs
with high aspect ratio, revealing the considerable influence of CNCs
morphology on the flow behavior of corresponding composite materials at all frequencies. The increase in G' and η^* with CNCs content
as well as aspect ratio can be attributed to the enhanced polymer-
filler interactions due to the formation of effective network junction
[84].

Modification of CNCs surface by surfactant coating, grafting and functionalization improves the interaction with polymer matrix and correspondingly exhibits significant impact on the rheological properties of resulting materials. For example; Littunen *et al.* [85] reported that the nanocomposite of poly(methyl methacrylate) (PMMA)-grafted-CNCs with PMMA exhibited high elasticity as well as low complex viscosity and tan δ as compared to non-grafted CNCs/PMMA composites. This was attributed to the enhanced degree of adhesion and interactions between grafted CNCs and polymer matrix.

6.6 Pros and Cons of CNCs: Surface Modification

The polar nature and high aggregation tendency of CNCs restrict their compatibility with non-polar media and hydrophobic polymer matrices. Such limitations can be overcome to a large extent by modifying the CNCs surface.

The surface modification can be carried out by (1) physical treatment like solvent extraction, stretching, calendaring, etc., (2) physico-chemical means such as use of corona and plasma discharge, γ-rays, UV bombardment, etc., and (3) chemical treatment (using either covalent or non-covalent method) depending on the polymer matrix. The selection criteria for the method to be used for surface modification lies in cost effectiveness, scalability, interfacial properties and suitability for processing conditions [86].

The most promising and widely used strategy is the chemical functionalization of CNCs surface [87]. Polymer coating and adsorption of surfactants on CNCs surface have also been studied to tune the compatibility with polymer matrices [88,89]. The physical methods do not change the surface properties of CNCs significantly, thus, resulting in lower degree of compatibility between CNCs and polymer matrix, whereas the physico-chemical methods are non-economical and make the approach unsuitable for large scale production.

The surface hydroxyl groups of CNCs provide opportunities to introduce different chemical functionalities on the surface. Generally, the surface functionalization of CNCs includes sulfonation, silylation, esterification, acetylation, carboxylation, oxidation, etherification, amidation, carbamation, cationization, nucleophilic substitution and grafting of polymer moieties on the surface [4,54,81,90-95]. Figure 6.7 represents different surface chemical modification methods.

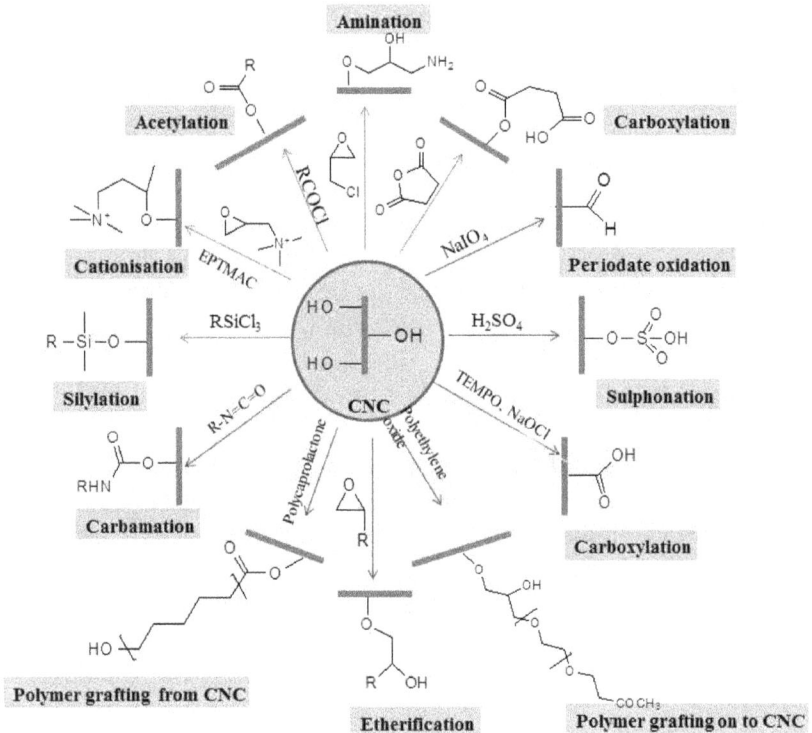

Figure 6.7 Schematic representation of the functionalization of CNCs surface.

The main objective of surface functionalization is to reduce the surface energy of CNCs by introducing positively or negatively charged groups and, thus, enhance their dispersion in the polymer matrices (hydrophilic/hydrophobic) by developing effective interfacial interactions. However, the major challenge lies in maintaining the integrity of CNCs and preserving their original morphology during chemical modification. The functionalized CNCs are potentially utilized in various applications such as films and nanocomposites, biomedicine, food, cosmetics, waste water treatment, energy sector, protective coatings, etc.

6.7 Processing of CNCs in Polymer Nanocomposites

Following are the two major techniques commonly used for processing of CNCs in the polymer matrices for developing functional hybrids and composites:

6.7.1 Casting-Evaporation Method

Wet processing methods, such as casting-evaporation, are the most extensively reported technique for the preparation of CNCs based nanocomposites. In this method, CNCs dispersion in a suitable processing medium is thoroughly mixed with polymer solution/dispersion, followed by casting and evaporation, thus, resulting in solid nanocomposite films.

Water is considered as the most suitable processing medium for CNCs to compound with polymers due to the formation of stable colloidal suspensions of CNCs. However, it also restricts the type of polymer matrix used. Other polar solvents, such as N,N dimethyl acetamide (DMAC) and N,N dimethyl formamide (DMF), can also be used as processing media for the preparation of CNCs based composite materials, by using solvent exchange procedure to obtain CNCs suspensions in appropriate media for mixing with suitable polymer solutions [96]. Alternatively, re-dispersion of lyophilized CNCs in organic media can be used for mixing with polymer solutions [97,98].

Moreover, modified CNCs (chemically functionalized or surfactant coated) exhibit effective dispersion in non-polar or hydrophobic media, thus, leading to uniform mixing with polymer solutions, as modification reduces the surface energy of CNCs and enhances their compatibility. For instance, it has been reported by Habibi and Dufresne [99] that the polycaprolactone (PCL) grafted CNCs showed good dispersion as compared to unmodified CNCs in the PCL matrix, thus, representing a promising approach for processing of nanocomposite materials.

Although casting-evaporation method preserves the dispersion state of CNCs in the liquid, however, it limits the number of polymer matrices that can be used. In addition, it is non-economical and non-industrial. Thus, industrial-scale nanocomposite processing techniques are needed for effective commercialization of these composites.

6.7.2 Melt Compounding

Extrusion and injection molding are particular methods of melt compounding for processing of thermoplastics. As melt compounding does not require any solvent, it is considered as a green technology for processing of polymer composites or blends and is economically/industrially more feasible. Several factors which are primarily considered while using this processing technique for composite preparation such as:

1. Suitable physical form of nanoparticles to be incorporated in the polymer melt without adversely affecting the intrinsic properties of the base polymer
2. Dispersion and aggregation of nanoparticles in the polymer melt
3. Surface functionalization of nanomaterials for better dispersion and strengthening of nanoparticle-polymer interface
4. Thermal stability of nanoparticles

In this process, dried nanomaterials are incorporated in the polymer melt and mixed by using internal mixers to produce uniformly compounded mixture of nanofiller and polymer matrix, followed by processing via film blowing, injection or compression molding, etc. However, the melt compounding of CNCs with polymer matrices for fabrication of composite materials has not been extensively studied due to aggregation and thermal issues related to CNCs.

In recent years, encouraging results have been reported for surface functionalized CNCs based composite materials. Menezes *et al.* [100] reported that CNCs grafted with organic acid chloride of varying aliphatic chain lengths were successfully extruded with LDPE at 160 °C without thermal decomposition. It was observed that the CNCs modified with longer acid chains showed good dispersion in the polymer matrix. In another study [101], poly(ethylene glycol) (PEG) grafted CNCs were coated with poly(ethylene oxide) (PEO) and subsequently extruded with polystyrene (PS) at elevated temperature (200 °C) to prepare nanocomposites with good dispersion and improved thermal stability.

The properties of CNCs based nanocomposites are significantly affected by the processing methods used to fabricate composite materials. Hajji *et al.* [102] evaluated the mechanical properties of CNCs reinforced latex copolymer prepared by three different processing methods, viz. (a) solution casting, (b) freeze drying followed by hot pressing and (c) freeze drying and extrusion followed by hot pressing. The composites fabricated by solution casting process displayed best mechanical performance as the method had less influence on the CNCs orientation and provided better dispersion in the polymer matrix without causing damage to the CNCs structure.

6.7.3 Other Processing Techniques

In addition to the earlier described methods, other techniques such as electro-spinning, layer by layer assembly and sol-gel processing have also been studied for the preparation of CNCs-polymer nanocomposites.

Electrospinning, a versatile method to prepare polymer nanofibers in the presence of electrostatic forces [103], has emerged as an alternative process to incorporate CNCs in polymer matrices such as PVA, PEO, poly(acrylic acid) (PAA), etc. Electrospun PCL nanofibers having nano-crystalline cellulose as reinforcement have been reported by Zoppe *et al.* [104]. The nanofibers exhibited improved modulus and strength when compared with neat PCL nanofibers which is attributed to the differences in the fiber diameter, CNCs loading and crystallization process. Mgalahas *et al.* [105] reported a core-in shell electrospinning technique to produce electrospun arrays of CNCs in which cellulose nanocrystals constituted the core component surrounded by cellulose shell.

Layer by layer assembly is a unique method for the preparation of nanocomposite films which involves alternate deposition of component layers with opposite charges via spin coating or solution dipping. This method has been commonly applied to prepare cellulose nanocrystals based multilayer composite films with different functionalities [106,107].

In addition to above described methods, *in-situ* polymerization of monomers in the presence of CNCs has also been reported to develop nanocomposite materials with uniformly distributed CNCs, exhibiting reduced moisture content and enhanced interaction with polymer matrices [108,109]. The major limitation of this method lies in the fact that it is only applicable to liquid phase polymerization, where liquid monomers are polymerized in the presence of nanocellulose filler.

A facile template method has been described by Capadona *et al.* [110] for processing of nanocomposites which involves the formation of a 3D template of CNCs through a sol-gel process (carried out via solvent exchange procedure). The template is subsequently filled with selective polymers by immersing the gel into the desired polymer solution followed by drying and shaping. Care must be taken in selecting the solvent for the polymer solution as it should mix well with the gel solvent and restrict the re-dispersion of cellulose nanocrystals in it.

6.8 Applications

Over the last few years, CNCs have been utilized in different industrial and non-industrial sectors. The applications of CNCs include adhesives, electrically conductive papers, perm-selective membranes, aerogels, Pickering emulsions, protein immobilization, antimicrobial agents, biosensors, green catalysts, drilling fluids, etc. [111]. Some of the industrial uses are briefly described in the sections below:

6.8.1 Packaging

Petroleum derived synthetic plastics are utilized in flexible packaging due to ease of handling, lightweight nature and cost effectiveness. About 50% of the total amount of plastics is used for packaging of disposable things, and 20% of this amount is utilized in food packaging per year [112].

Although synthetic plastics driven from petrochemicals have several advantages of optimal mechanical and barrier properties, low cost, easy availability, etc., however, the major drawback is their non-biodegradable nature, thus, generating significant municipal solid wastes [113]. Thus, plastic industries are focused on developing biodegradable plastics meeting the requirement of packaging materials. Use of biopolymers in the recent past has seen an exponential growth, and such polymers are generally categorized as natural and synthetic biopolymers [17]. In most of the cases, poor mechanical and barrier properties of these biopolymers inhibit their wide spread utilization in packaging applications. Various attempts have been made to resolve the limitation, and the use of fillers in polymer matrices is observed to be the most effective remedy. Growing know-how in nanotechnology has led to the development of bio-nanocomposites, as one of the promising way, by incorporating suitable nanofillers in biopolymers [9,114]. Among various nanofillers, cellulose nanocrystals owing to their excellent features are promising candidates for packaging applications and have been utilized in many biopolymers such as starch, chitosan, PCL, poly(β-hydroxyoctanoate), xylan, poly(vinyl chloride), etc. [115-120].

Khan *et al.* [116] reported chitosan films reinforced with CNCs by solvent casting method to evaluate the effect of CNCs on the mechanical, barrier and thermal properties of the resulting films. With only 5% loading of CNCs, the tensile strength and modulus were increased by 26% and 87% respectively, whereas water vapor permeability reduced by 27%. However, addition of CNCs had no significant effect on the thermal property of the films. The enhancement in mechanical and barrier properties was attributed to the interfacial interactions and formation of percolation network.

In another study, barrier properties of CNCs incorporated PVA films (cross-linked with PAA) were investigated by Paralikar *et al.* [69]. The films with 10% CNCs showed reduced WVTR due to high crystallinity and uniform dispersion of CNCs in the matrix. However, agglomeration was observed for more than 10% loading, revealing inferior properties of the films. The effect of CNCs surface modification was also evaluated, where improvement in the film properties was observed by using carboxylated CNCs as compared to unmodified CNCs. Nanocomposite films of poly(3-hydroxybutyrate-co-3-hydroxyvalerate) (PHVB) and CNCs were fabricated for biodegradable food packaging application [121]. The resulting films displayed superior mechanical, barrier and migration properties compared to neat PHVB films. The optimal values were obtained at 20 wt% loading of CNCs [121].

Fortunati *et al.* [122] reported hybrid nanocomposite films of PLA, prepared via melt extrusion process, by incorporating surfactant modified CNCs and silver nanoparticles (Ag NPs). The resulting films exhibited enhanced mechanical, barrier and thermal properties owing to the presence of CNCs, while the antimicrobial response of the films was attributed to the Ag NPs. The surfactant coating on CNCs surface improved their dispersion in the PLA matrix, thus, resulting in transparent films with superior properties. The optimal properties were observed for 5% CNCs and 1% Ag NPs loadings.

6.8.2 Waste Water Treatment

Increasing population and human activities create a serious water pollution problem. The discharge from the industries contains various contaminants including heavy metal ions, dyes, organic and inorganic particles, pharmaceuticals, pesticides, other hazardous materials and biological pollutants such as bacteria, fungi, algae, etc. As only 0.27% of the total fresh water present on the earth surface is useful, it is, therefore, necessary to improve the water quality by removing the impurities [123]. Many techniques have been developed for water treatment, however, adsorption is observed to be the most effective and environment friendly method [124,125].

Among the conventional adsorbents, activated carbon is widely used for water remediation. However, costly and energy-intensive production of these adsorbents escalate the need for the development of alternative cost-effective adsorbents [124]. In this respect, intensive research efforts have been focused on the use of nano-crystalline cellulose for waste water treatment as adsorbent or water filtration membranes.

The adsorption efficiency of CNCs towards different pollutants is governed by surface area as well as nature and charge density of the functional groups present on the CNCs surface. Several literature studies have revealed that the CNCs obtained after usual sulfuric acid hydrolysis exhibit good adsorption efficiency towards pollutants. The anionic sulfate groups introduced at CNCs surface during hydrolysis induce adsorption of cationic species. Liu *et al.* [126] reported that the adsorption efficiency of CNCs towards Ag^+ was almost twice than CNFs due to the presence of sulfate groups. Overall, CNCs have been investigated in different studies as bio-adsorbent for the removal of various toxic metal ions, such as Cd^{2+}, Pb^{2+}, Ni^{2+}, Ag^+, Cu^{2+} and Fe^{3+}, from water bodies [127-129,130].

The surface functionalization of CNCs with different anionic groups such as phosphate and carboxylate moieties was observed to enhance the adsorption capacity and selectivity of CNCs for the uptake of cationic toxic addendums [131-133]. Furthermore, positively charged CNCs have also been studied for the sorption of negatively charged metal ions like arsenates, chromates, vanadates, etc. [81,134].

Grafting CNCs surface with polymers containing reactive functional group is a very promising approach to enhance adsorption capacity for various species [130,135]. However, difficulty lies in their immobilization on CNCs surface due to nano-dimensions, high surface area and steric hindrance.

Apart from heavy metal ions, organic dyes produced in large quantities in textile industries cause threat to human health. Batmaz *et al.* [123] reported that the adsorption and removal of the cationic dye, i.e. methylene blue (MB), increased with the use of carboxylated CNCs, obtained through TEMPO mediated oxidation of pristine CNCs (produced via sulfuric acid hydrolysis of micro-crystalline cellulose). It was reported that the CNCs with high content of COOH groups (2.1 mmol/g) showed higher uptake value of 769 mg dye/g CNCs at pH=9 for the mentioned dye as compared to sulfonic acid functionalized CNCs (118 mg/g at same pH).

In another study [136], carboxyl group containing CNCs, obtained in a single step through oxidation of ammonium persulfate, exhibited an adsorption of 101 mg/g for methylene blue at neutral pH. The functionalized CNCs removed more than 90% of the dye in seven cycles of sorption and desorption using ethanol as eluent.

Moreover, removal of negatively charged dyes, like Congo red, Orange II, etc., via adsorption through CNCs containing positively charged surface groups has also been reported [137-139]. Amine containing positively charged CNCs were prepared by Jin *et al.* [140] through successive sodium periodate oxidation of pristine CNCs, followed by reaction with ethylene diamine. The developed nanomaterial exhibited a maximum uptake of 556 mg/g for acid red GR dye.

CNCs incorporated nanocomposites have also been demonstrated for the adsorption of MB. For instance, Mohammed *et al.* [141] prepared CNCs incorporated alginate beads by ionic crosslinking of alginate chains through $CaCl_2$. The presence of sulfate groups on CNCs and carboxylate groups on alginate contributed to the observed MB adsorption of 256 mg/g. The developed materials could be reused without loss in adsorption capacity even after five cycles of sorption-desorption by using 1:1 mixture of HCl and ethanol as eluent.

6.8.3 Energy Storage Devices

Modernization and technological growth have increased the demand for energy in different sectors. Thus, the development of suitable electronic and energy storage devices, such as lithium ion batteries, supercapacitors and fuel cells for various portable, stationary and automotive applications, has become necessary. Nevertheless, the raw materials used for the fabrication of energy devices are mostly based on ceramics, metals, plastics and other petrochemical derived products, which raises concerns about sustainability, environmental issues, among others.

In this respect, nanosized cellulose derivatives, i.e. CNCs, CNFs and BC, have promising potential to mitigate the negative impact of the conventional materials and have received increasing attention in the realm of organic electronics and energy storage. Several literature studies have reported on the use of CNCs based nanocomposites as substrates for flexible printed circuits, conductive thin films, separators in lithium ion batteries and electrode materials in supercapacitors. A few of these studies are briefly discussed in the sections below

Nanocellulose Derived Conductive Nanocomposite Films

Conductivity is the prime requirement for the core components of energy devices, thus, electrically conductive substrates are in significant demand. Unfortunately, nanocelluloses, in spite of having excellent inherent properties, are electrically non-conductive and cannot be directly used for the fabrication of energy devices. However, nanocellulose based conductive materials can be developed by combining with conducting polymers, conducting carbon materials, metallic particles, etc. [142].

The resulting composites combine the advantages of both CNCs and conductive materials, which can meet the requirements for energy applications. For instance, inherently conducting polymers exhibit excellent chemical and electronic properties, however, have inferior mechanical properties [143]. To overcome this limitation, efforts have been made to prepare CNCs reinforced conductive nanocomposites of such conducting polymers, especially polyaniline (PANI) and polypyrrole (Ppy), for energy applications. The conducting nanocomposites can be fabricated either by coating the CNCs surface with conducting polymers or by integrating them in the polymer matrix via *in-situ* polymerization. The presence of CNCs is observed to lowering the percolation threshold and enhance mechanical performance [144].

Hamad and Atifi [145] prepared conductive films of CNCs containing PANI by *in-situ* or emulsion polymerization. The films with 2:1 weight ratio of CNCs:PANI were reported to have strength and conductivity of 9.7 $\pm$ 1.8 MPa and 1.88 x 10^{-2} S/cm respectively. Thus, the incorporation of CNCs provided flexibility, conductivity and strength to the resulting composites.

Flexible laminated membranes of alternately deposited 1D CNCs and 2-D graphene oxide (GO) nano-sheets with excellent mechanical properties and electrical conductivity were reported by Xiong *et al.* [146] via layer by layer technique. To enhance the interfacial interaction between the two layers, CNCs were modified with cationic polyetherimide (PEI). The resulting membranes with 60 μm thickness exhibited conductivity of 5000 S/m with specific strength and stiffness of 382 MPa/gcm^3 and 100 GPa/gcm^3 respectively, thus, making them suitable for developing electrical devices.

In another study, stable and flexible composite films of Ppy coated CNCs were reported with electrical conductivity values ranging from 10^{-3} to 10 S/cm [147]. Highly conductive nanocellulose composites have also been prepared in other studies by incorporating carbon nanomaterials such as carbon nanotubes (CNTs) [148], GO [149] and r-GO [150] via surface coating or blending. The composites exhibited excellent strength and higher conductivity as compared to composites containing CNCs.

A ternary nanocomposite of multi-walled carbon nanotubes (MWCNTs)/GO/CNCs was demonstrated by Tang *et al.* [151]. The conductive carbon nanomaterials were introduced into cellulose nano-paper via blending process in which GO was used as a dispersant to improve the dispersion of MWCNTs in water media, while CNCs acted as a binder for the MWCNTs/GO nanocomposites. The hybrid nano-paper exhibited a conductivity of 8.92 S cm^{-1} with high mechanical strength.

Flexible Printed Circuit Boards

Flexible printed electronics is promising field of research that combines conductive material formulation and printing processes for the manufacture of electronic components. The market value of this industry is estimated to reach USD 73 billion by 2025 [143]. Till date, the use of nanocellulose in this field is still very limited as compared to other applications. The available literature reviews and patents demonstrate that nanocellulose has the potential of application in all aspects of flexible printed electronics including conductive inks, substrates and electronic devices [143].

Flexible and transparent substrates constitute major breakthrough in the development of printed electronics. This field requires thermally stable, flexible, smooth and non-porous substrates, which can withstand high temperature and avoid the probable breakage of conductive paths leading to reduced electrical conductivity. Nanocellulose, being biodegradable, fulfills the requirements of substrate materials for use in printed electronics, thus, providing an alternative to plastic and glass substrates [152]. CNFs based nanocellulose substrates are mostly applied for printed electronics rather than CNCs owing to more brittle nature of CNCs films as compared to CNFs.

Surface roughness and porosity are important properties of substrates which impact print quality by affecting ink penetration and ink-film continuity. Though higher roughness reduces conductivity, it can also increase ink anchoring. Significant research efforts are underway for the development of new and innovative ways to circumvent the issues arising from the high levels of porosity and roughness inherent to paper substrates. Several methods such as internal sizing, calendaring and use of coatings have been applied to develop paper based substrates with improved qualities [153-155]. To reduce surface roughness, use of cellulose nanocrystals (CNCs) has been reported by Hoeng *et al.* [156], where a cardboard surface showed reduction in roughness from 66 nm to 3.6 nm, after coating with TEMPO (2,2,6,6-Tetramethylpiperidine 1-oxyl) oxidized CNCs.

CNCs in Supercapacitors

Supercapacitors are a promising approach to meet the increasing energy demand as these exhibit high energy density, high power density, rapid charging-discharging rate, long life and low maintenance cost. High surface area and conductivity are major factors affecting the performance of supercapacitors. However, materials with highly porous structure, large specific area, conductivity and high mechanical strength are required for electrodes of supercapacitors. CNCs with high surface area facilitate high charge storage and have significant potential in the field of supercapacitors.

Liew *et al.* [157] synthesized nanocomposites of CNCs incorporated PPy through electro-deposition method for stable supercapacitors. In this process, CNCs surface was first modified by introducing carboxyl functional groups through TEMPO mediated oxidation, which facilitated the subsequent electro-deposition of PPy on the surface to fabricate CNCs/PPy nanocomposites. A symmetrical supercapacitor having two CNC/PPy nanocomposite based electrodes was fabricated and examined at operating voltage of 1 V. The energy density of the resulting supercapacitor was reported to be 8.34 W h/Kg at 1 mA/cm^2 current density, and it retained 47% of its initial capacitance after 50,000 charging-discharging cycles. The superior cycling stability of the supercapacitor was attributed to the strength provided by CNCs as well as the porous structure of the nanocomposite which facilitated the movement of ions and solvent molecules during charge/discharge cycles, making the structure more stable.

Flexible and lightweight aerogels of modified CNCs with good mechanical strength and shape recovery ability were developed by Yang *et al.* [158] for supercapacitor systems. The aerogels with hydrazone crosslinks were prepared by chemical crosslinking between CNCs containing aldehyde functional groups and hydrazine modified CNCs. The aerogels were further loaded with three different types of capacitive nanoparticles (CNP), i.e. Ppy nanofibers (Ppy-NF), Ppy-coated CNTs (Ppy-CNT) and manganese dioxide nanoparticles (MnO_2-NP) at weight ratio of 1.5:1 for CNP:CNCs, to fabricate hybrid aerogels. The aerogels offered high surface area of active nanoparticles to promote charge storage with excellent capacitance retention of 84.2%, 61.7% and 92.3% after 2000 cycles for Ppy-NF, Ppy-CNT and MnO_2-NP respectively. The aerogels also displayed low internal resistance owing to highly porous nature of the material. Shape recovery experiments, carried out by compressing the hybrid aerogels in both air and Na_2SO_4 aqueous electrolyte, indicated that each aerogel possessed a height recovery rate of 0.05 cm/s. Ppy-NF, Ppy-CNT and MnO_2-NP containing aerogels maintained 90±3%, 77±4% and 83±3% of their original heights respectively after 400 compression cycles at 80% strain, which confirmed the suitability of developed aerogels for flexible electronic applications.

Lithium Ion Batteries (LIBs)

In addition to supercapacitors, CNCs have been employed as separators for LIBs. LIBs are the most promising power source in wide range of applications from portable devices to electric vehicles. Rapid industrial growth necessitates LIBs to have higher energy and power density, lower manufacturing cost and eco-friendly nature. To date, nanocellulose has been employed in all components of LIBs, i.e. anode, cathode, separator and electrolyte.

Separator plays an important role in preventing direct contact between the two electrodes. At the same time, it serves as an electrolyte reservoir to transport the ions between the electrodes for maintaining conductivity. The structure and properties of the separators have a significant influence on the performance of LIBs. Thus, essential requirement for a separator material include good wettability for electrolytes, strong mechanical strength, porous structure having appropriate pore size, appropriate thickness, good electrochemical and thermal stability, etc. [159].

In the recent years, CNCs reinforced solid flexible polymeric films have gained attention as separator materials for fulfilling the above mentioned requirements [160]. For instance, Samir *et al.* [161,162] evaluated CNCs reinforced PEO based polymeric membranes for high conductivity, combined with excellent electrochemical, thermal and mechanical stability. Among the different ways of fabricating nanocomposite films, electrospun nonwoven mats of CNCs based nanocomposites with appropriate porosity, flexibility and surface area are receiving increasing attention as separator materials [159].

Lalia and co-workers [163] developed nanocomposite films of CNCs reinforced poly(vinylidene fluoride-co-hexafluoropropylene) (PVDF-HFP) by electrospinning, to be used as separator for Li-ion batteries. The films were investigated for their thermomechanical and electrochemical properties. With addition of only 2% CNCs (optimum concentration) in the composite mat, an increment of 15% in storage modulus as well as 63% enhanced retention of 1 M (LiTFSI/BMPyrTFSI) electrolyte were observed, as compared to the mats without CNCs. The resulting electrolyte membranes accomplished high conductivity (i.e. $4x10^{-4}$ S/cm at 30°C) and good electrochemical stability up to 5 V, thus, making them viable for LIBs.

In another study, porous and dense CNCs incorporated poly(vinylidene fluoride) (PVDF) nanocomposite membranes were investigated as separator material in LIBs [164]. The addition of CNCs induced significant improvement in Young's modulus, in both dense and macro-porous membranes. Moreover, the presence of CNCs facilitated the formation of polar β phase of PVDF in the macro-porous membranes, which preferentially improved the wettability of the separators. Although conductivity was observed to slightly decrease with the incorporation of CNCs owing to decrease in porosity and lower pore interconnection, however, the developed materials outperformed commercial grade polyolefin based separator (Celgard®2400), thus, confirming their potential use in lithium ion batteries.

6.8.4 Biomedical Engineering

The unique features of CNCs make them a promising material for biomedical applications. A large number of literature studies have been performed to explore the potential of cellulose nanocrystals in this field, a few of which are reviews in the following sections.

Tissue Engineering

Tissue engineering is an interdisciplinary field which deals with the substitution of diseased or malfunctioning organs with natural, synthetic or semi-synthetic counterpart. The substitutes are designed to mimic the natural organization and function of the target tissue.

CNCs along with cellulose acetate have been electrospun for vascular regeneration, which exhibited reduction in thrombogenecity and increase in tensile strength [165]. Li *et al.* [166] developed CNCs reinforced collagen films and observed enhancement in mechanical properties as well as swelling tendency. The nanocomposites also exhibited good bio-adhesion of cells, thus, confirming the suitability for wound dressing application.

PVA/CNC nanofiber mats, preferably suited for skin tissue engineering application, have been developed by Lee *et al.* [167]. The mats demonstrated 35% and 45% increment in tensile strength and elastic modulus respectively as compared to the isotropic PVA mat. Also, the mats possessed nanofibers of comparatively lower diameter which can impart cell adhesion sites and stretchability. Reuda and co-workers [168] developed nanocomposites of CNCs and polyurethane (PU) with improved ductility and toughness at optimal CNCs concentration of 5%. The resulting nanocomposites did not show any toxicity towards L-929 fibroblast cells, thus, highlighting their potential for wound healing and drug loaded implant coatings. In another study, CNCs were electrospun with cellulose to fabricate uniaxially aligned cellulose nanofibers. The resulting fiber matrix with 20% CNCs concentration showed 101.7% and 171.6% increase in tensile strength and elastic modulus respectively, as compared to the pure cellulose matrix. The developed fiber matrix has potential of application in blood vessels, tendon, skin, nerve regeneration, etc. [169].

Kumar *et al.* [170] synthesized PVA/nano-hydroxyapatite (n-HA)/CNCs based porous composite scaffold with improved mechanical and bioactive properties as compared to native PVA scaffold. The authors also observed an increase in the compressive strength as well as decreased porosity on varying the CNCs concentration. Zhou *et al.* [171] grafted maleic anhydride on PLA for interfacial adhesion of CNCs with the polymer matrix. The resulting matrix was used for bone tissue engineering applications and showed increased stability during biodegradation and increment in mechanical properties.

CNCs-pectin and CNCs-carboxy methyl cellulose demonstrated tunable swelling properties which can possibly provoke controlled drug release tendency for water soluble drugs, thus, confirming the application as drug eluting scaffold [172].

Hydrogels

Hydrogels are macromolecular polymer gels formed by physically or chemically crosslinked network of hydrophilic monomers. Their applications in biomedical field include contact lenses, transdermal ionophoresis gels, wound dressing gels for treating burns, gastric retentive devices, etc. CNCs with high aspect ratio, crosslinking affinity and ease of chemical modification exhibit potential as reinforcing filler in hydrogel matrices for different medical applications.

Domingues *et al.* [173] developed injectable hydrogels of hyaluronic acid and CNCs for tissue engineering applications. Stiffer hydrogels with compact network structure were observed on incorporation of CNCs in the matrix. The hydrogels also exhibited cell supportive properties when assessed towards adipose stem cells.

Cellulose nanocrystals incorporated polyacrylamide hydrogels were developed by Yang *et al.* [174] via free radical polymerization. Bonding between the chains of polyacrylamide and CNCs resulted in an elastic network formation which led to the conformational rearrangements leading to increased mechanical and viscoelastic properties. The developed hydrogels can be used for the fabrication of artificial muscles, ligaments, etc.

Yang *et al.* [175] synthesized injectable hydrogels of CMC, dextran and CNCs. Crosslinked hydrogels were reported to possess significantly higher elastic moduli (>140% increase at peak strength) compared to native hydrogels, without changing the pore structure and with negligible cytotoxicity.

Drug Delivery System

Drug delivery involves strategies to transport a pharmaceutical compound inside the body for curing the diseased state. Drug delivery ranges from conventional methods like oral (tablets, capsules, elixirs, etc.), intramuscular and cutaneous delivery (via injection) to vehicle-assisted (nano-carriers, dermal patches, hydrogels, controlled release formulations, etc.). CNCs serve as nano-carriers which be attached with drug molecules or as a filler for drug adsorbing matrix for tuning its controlled release. CNCs possess exceptional characteristic of being phagocytized by the cells, thus, favoring cellular uptake of CNCs-drug conjugate by the target cells during therapy. Chemical modification of CNCs paves the way for the attachment of hydrophobic or non-ionizable drugs on the surface [91].

Wang *et al.* [176] developed polyelectrolyte macro-ion complex (PMC) of chitosan and cellulose nanocrystals for application in oral drug delivery. PMC possessed a net positive charge and a size <10 μm, which were highly favorable parameters for attaching and delivering negatively charged drugs. Jackson *et al.* [177] modified CNCs surface with CTAB to bind hydrophobic drugs paclitaxel (PTX), docetaxel (DTX) and etoposide ETOP. The increase in the zeta potential of particles from -55 to 0 mV resulted in an enhancement of the binding affinity for hydrophobic drugs. Controlled release of the drugs was exhibited over several days, which revealed the therapeutic potential of developed materials to KU-7 bladder cancer cells.

6.9 Conclusion

Nanocellulose has confirmed its suitability for a plethora of applications till date due to its unique features of high aspect ratio, large surface area and excellent mechanical properties along with biocompatibility, non-toxicity, chemical resistivity and abundant availability. The market potential for nanocellulose in the fields of food, energy and biomedicine is immense. With the ongoing technological advancements, many commercial applications of nanocellulose can be envisaged in the near future.

Acknowledgement

The authors are grateful to University Grant Commission, India for financial support during the study.

References

1. Tang, J., Sisler, J., Grishkewich, N., and Tam, K. C. (2017) Functionalization of cellulose nanocrystals for advanced applications. *Journal of Colloid and Interface Science,* **494**, 397-409.
2. Grishkewich, N., Mohammed, N., Tang, J., and Tam, K. C. (2017) Recent advances in the application of cellulose nanocrystals. *Current Opinion in Colloid and Interface Science,* **29**, 32-45.
3. Trache, D., Hussin, M. H., Haafiz, M. K. M., and Thakur, V. K. (2017) Recent progress in cellulose nanocrystals: sources and production. *Nanoscale,* **9**(5), 1763-1786.
4. Nascimento, D. M., Nunes, Y. L., Figueirêdo, M. C. B., De Azeredo, H. M. C., Aouada, F. A., Feitosa, J. P. A., Rosa, M. F., and Dufresne, A.

(2018) Nanocellulose nanocomposite hydrogels: technological and environmental issues. *Green Chemistry*, **20**(11), 2428-2448.

5. Islam, M. S., Chen, L., Sisler, J., and Tam, K. C. (2018) Cellulose nanocrystal (CNC)-inorganic hybrid systems: synthesis, properties and applications. *Journal of Materials Chemistry B*, **6**(6), 864-883.

6. Trache, D., Hussin, M. H., Hui Chuin, C. T., Sabar, S., Fazita, M. R. N., Taiwo, O. F. A., Hassan, T. M., and Haafiz, M. K. M. (2016) Microcrystalline cellulose: Isolation, characterization and biocomposites application - A review. *International Journal of Biological Macromolecules*, **93**, 789-804.

7. Bano, S., and Negi, Y. S. (2017) Studies on cellulose nanocrystals isolated from groundnut shells. *Carbohydrate Polymers*, **157**, 1041-1049.

8. Klemm, D., Heublein, B., Fink, H. P., and Bohn, A. (2005) Cellulose: Fascinating biopolymer and sustainable raw material. *Angewandte Chemie - International Edition*, **44**(22), 3358-3393.

9. Bharimalla, A. K., Deshmukh, S. P., Vigneshwaran, N., Patil, P. G., and Prasad, V. (2017) Nanocellulose-polymer composites for applications in food packaging: current status, future prospects and challenges. *Polymer-Plastics Technology and Engineering*, **56**(8), 805-823.

10. de Souza Lima, M. M., and Borsali, R. (2004) Rod like cellulose microcrystals: structure, properties and applications. *Macromolecular Rapid communications*, **25**, 771-787.

11. Kobayashi, S., Kashiwa, K., Shimada, J., Kawasaki, T., and Shoda, S. (1992) The first in vitro synthesis of cellulose via nonbiosynthetic path catalyzed by cellulase. *Macromolecular Chemistry*, **54-55**, 509-518.

12. Roy, D., Semsarilar, M., Guthrie, J. T., and Perrier, S. (2009) Cellulose modification by polymer grafting: A review. *Chemical Society Reviews*, **38**(7), 2046-2064.

13. Naduparambath, S., Jinitha, T. V., Shaniba, V., Sreejith, M. P., Balan, A. K., and Purushothaman, E. (2018) Isolation and characterisation of cellulose nanocrystals from sago seed shells. *Carbohydrate Polymers*, **180**, 13-20.

14. Rowland, S. P., and Roberts, E. J. (1972) The nature of accessible surfaces in the microstructure of cotton cellulose. *Journal of Polymer Science, Part A: Polymer Chemistry*, **10**(8), 2447-2461.

15. Habibi, Y., Lucia, L. A., and Rojas, O. J. (2010) Cellulose nanocrystals : chemistry, self-assembly, and applications. *Chemical Reviews*, **110**, 3479-3500.

16. Chakraborty, S., Chowdhury, S., and Das Saha, P. (2011) Adsorption of crystal violet from aqueous solution onto NaOH-modified rice husk. *Carbohydrate Polymers*, **86**(4), 1533-1541.

17. George, J., and Sabapathi, S. N. (2015) Cellulose nanocrystals : Syn-

thesis, functional properties, and applications. *Nanotechnology, Science and Applications,* **8**, 45-54.

18. French, A. D., and Santiago Cintrón, M. (2013) Cellulose polymorphy, crystallite size, and the segal crystallinity index. *Cellulose,* **20**(1), 583-588.

19. Atalla, R. H., and VanderHart, D. L. (2011) Native cellulose : A composite of two distinct crystalline forms. *Science,* **223**(4633), 283-285.

20. Nishiyama, Y., Sugiyama, J., Chanzy, H., and Langan, P. (2002) Crystal structure and hydrogen bonding system in cellulose $I\alpha$ from synchrotron x-ray and neutron fiber diffraction. *Journal of American Chemical Society,* **124**, 9074-9082.

21. Debzi, E. M., Chanzy, H., Sugiyama, J., Tekely, P., and Excoffier, G. (1991) The $I\alpha \rightarrow I\beta$ transformation of highly crystalline cellulose by annealing in various mediums. *Macromolecules,* **24**(26), 6816-6822.

22. Kim, N. H., Imai, T., Wada, M., and Sugiyama, J. (2006) Molecular directionality in cellulose polymorphs. *Biomacromolecules,* **7**(1), 274-280.

23. Wada, M., Heux, L., and Sugiyama, J. (2004) Polymorphism of cellulose I family: reinvestigation of cellulose IV_I. *Biomacromolecules,* **5**(4), 1385-1391.

24. Gardiner, E. S., and Sarko, A. (1985) Packing analysis of carbohydrates and polysaccharides. The crystal structures of celluloses IV I and IV II. *Canadian Journal of Chemistry,* **63**(1), 173-180.

25. Mariano, M., El Kissi, N., and Dufresne, A. (2014) Cellulose nanocrystals and related nanocomposites: review of some properties and challenges. *Journal of Polymer Science, Part B: Polymer Physics,* **52**(12), 791-806.

26. Nascimento, D. M., Nunes, Y. L., Figueirêdo, M. C. B., De Azeredo, H. M. C., Aouada, F. A., Feitosa, J. P. A., Rosa, M. F., and Dufresne, A. (2018) Nanocellulose nanocomposite hydrogels: technological and environmental issues. *Green Chemistry,* **20**(11), 2428-2448.

27. El Achaby, M., Kassab, Z., Barakat, A., and Aboulkas, A. (2018) Alfa fibers as viable sustainable source for cellulose nanocrystals extraction: application for improving the tensile properties of biopolymer nanocomposite films. *Industrial Crops and Products,* **112**, 499-510.

28. Seabra, A. B., Bernardes, J. S., Fávaro, W. J., Paula, A. J., and Durán, N. (2018) Cellulose nanocrystals as carriers in medicine and their toxicities: A review. *Carbohydrate Polymers,* **181**, 514-527.

29. Shi, Z., Phillips, G. O., and Yang, G. (2013) Nanocellulose electroconductive composites. *Nanoscale,* **5**(8), 3194-3201.

30. Pääkkö, M., Vapaavuori, J., Silvennoinen, R., Kosonen, H., Ankerfors, M., Lindström, T., Berglund, L. A., and Ikkala, O. (2008) Long and

entangled native cellulose I nanofibers allow flexible aerogels and hierarchically porous templates for functionalities. *Soft Matter,* **4**(12), 2492-2499.

31. Saito, T., and Isogai, A. (2006) Introduction of aldehyde groups on surfaces of native cellulose fibers by TEMPO-mediated oxidation. *Colloids and Surfaces A: Physicochemical and Engineering Aspects,* **289**(1-3), 219-225.

32. Campano, C., Balea, A., Blanco, A and Negro, C. (2016) Enhancement of the fermentation process and properties of bacterial cellulose: A Review. *Cellulose,* **23**(1), 57-91.

33. Lelekakis, N., Wijaya, J., Martin, D., and Susa, D. (2014) The effect of acid accumulation in power-transformer oil on the aging rate of paper insulation. *IEEE Electrical Insulation Magazine,* **30**(3), 19-26.

34. Shelton, M. C., Us, T. N., Posey-dowty, J. D., Rios Perdomo, L. G., Dixon, D. W., Lucas, P. L., Wilson, A. K., Walker, K. R., and Lawniczak, J. E., Foulk, R. G., Phan, H. D., and Freeman, C. C. (2009) Low Molecular Weight Cellulose Mixed Esters and Their Use as Low Viscosity Binders and Modifiers in Coating Compositions, patent US7585905B2.

35. Rånby, B. G. (1951) Fibrous macromolecular systems. cellulose and muscle. The colloidal properties of cellulose micelles. *Discussions of the Faraday Society,* **11**(111), 158-164.

36. Emerton, H. W., Wrist, P. E., Sikorski, J., and Woods, H. J. (1952) Electron-microscopy of degraded cellulose fibres. *Journal of the Textile Institute Transactions,* **43**(11), T563-T564.

37. Revol, J. F., Bradford, H., Giasson, J., Marchessault, R. H., Gray, D. G. (1992) Helicoidal self-ordering of cellulose microfibrils in aqueous suspension. *International Journal of Biological Macromolecules,* **14**(3), 170-172.

38. Favier, V., Chanzy, H., and Cavaillé, J. Y. (1995) polymer nanocomposites reinforced by cellulose whiskers. *Macromolecules,* **28**(18), 6365-6367.

39. Mandal, A., and Chakrabarty, D. (2011) Isolation of nanocellulose from waste sugarcane bagasse (SCB) and its characterization. *Carbohydrate Polymers,* **86**(3), 1291-1299.

40. Johar, N., Ahmad, I., and Dufresne, A. (2012) Extraction, preparation and characterization of cellulose fibres and nanocrystals from rice husk. *Industrial Crops and Products,* **37**(1), 93-99.

41. Santos, R. M. dos, Flauzino Neto, W. P., Silvério, H. A., Martins, D. F., Dantas, N. O., and Pasquini, D. (2013) cellulose nanocrystals from pineapple leaf, a new approach for the reuse of this agro-waste. *Industrial Crops and Products,* **50**, 707-714.

42. Lu, P., and Hsieh, Y. Lo. (2012) Cellulose isolation and core-shell nanostructures of cellulose nanocrystals from chardonnay grape skins. *Carbohydrate Polymers,* **87**(4), 2546-2553.

43. Kallel, F., Bettaieb, F., Khiari, R., García, A., Bras, J., and Chaabouni, S. E. (2016) Isolation and structural characterization of cellulose nanocrystals extracted from garlic straw residues. *Industrial Crops and Products*, **87**, 287-296.

44. Nandi, S., and Guha, P. (2018) A review on preparation and properties of cellulose nanocrystal-incorporated natural biopolymer. *Journal of Packaging Technology and Research*, **2**(2), 149-166.

45. Kumar, A., Negi, Y. S., Choudhary, V., and Bhardwaj, N. K. (2014) Characterization of cellulose nanocrystals produced by acid-hydrolysis from sugarcane bagasse as agro-waste. *Journal of Materials Physics and Chemistry*, **2**(1), 1-8.

46. Sun, J. X., Sun, X. F., Sun, R. C., and Su, Y. Q. (2004) Fractional extraction and structural characterization of sugarcane bagasse hemicelluloses. *Carbohydrate Polymers*, **56**(2), 195-204.

47. Rosa, S. M. L., Rehman, N., De Miranda, M. I. G., Nachtigall, S. M. B., and Bica, C. I. D. (2012) Chlorine-free extraction of cellulose from rice husk and whisker isolation. *Carbohydrate Polymers*, **87**(2), 1131-1138.

48. Beck-Candanedo, S., Roman, M., and Gray, D. G. (2005) Effect of reaction conditions on the properties and behavior of wood cellulose nanocrystal suspensions. *Biomacromolecules*, **6**(2), 1048-1054.

49. Kargarzadeh, H., Ahmad, I., Abdullah, I., Dufresne, A., Zainudin, S. Y., and Sheltami, R. M. (2012) Effects of hydrolysis conditions on the morphology, crystallinity, and thermal stability of cellulose nanocrystals extracted from kenaf bast fibers. *Cellulose*, **19**(3), 855-866.

50. Roman, M., and Winter, W. T. (2004) Effect of sulphate groups from sulphuric acid hydrolysis on the thermal degradation behaviour of bacterial cellulose. *Biomacromolecules*, **5**, 1671-1677.

51. Camarero Espinosa, S., Kuhnt, T., Foster, E. J., and Weder, C. (2013) Isolation of thermally stable cellulose nanocrystals by phosphoric acid hydrolysis. *Biomacromolecules*, **14**(4), 1223-1230.

52. Sadeghifar, H., Filpponen, I., Clarke, S. P., Brougham, D. F., and Argyropoulos, D. S. (2011) Production of cellulose nanocrystals using hydrobromic acid and click reactions on their surface. *Journal of Materials Science*, **46**(22), 7344-7355.

53. Nishino, T., Matsuda, I., and Hirao, K. (2004) All-cellulose composite. *Macromolecules*, **37**(20), 7683-7687.

54. Moon, R. J., Martini, A., Nairn, J., Simonsen, J., and Youngblood, J. (2011) Cellulose nanomaterials review: structure, properties and nanocomposites. *Chemical Society Reviews*, **40**, 3941-3994.

55. Lu, P., and Hsieh, Y. Lo. (2010) Preparation and properties of cellulose nanocrystals: rods, spheres, and network. *Carbohydrate Polymers*, **82**(2), 329-336.

56. Jiang, B., Liu, C., Zhang, C., Wang, B., and Wang, Z. (2007) The effect

of non-symmetric distribution of fiber orientation and aspect ratio on elastic properties of composites. *Composites Part B: Engineering,* **38**(1), 24-34.

57. Mutjé, P., Lòpez, A., Vallejos, M. E., López, J. P., and Vilaseca, F. (2007) Full exploitation of cannabis sativa as reinforcement/filler of thermoplastic composite materials. *Composites Part A: Applied Science and Manufacturing,* **38**(2), 369-377.

58. Heux, L., Chauve, G., and Bonini, C. (2000) Nonflocculating and chiral-nematic self-ordering of cellulose microcrystals suspensions in nonpolar solvents. *Langmuir,* **16**(21), 8210-8212.

59. Elazzouzi-Hafraoui, S., Putaux, J. L., and Heux, L., (2009) Self-assembling and chiral nematic properties of organophilic cellulose nanocrystals. *Journal of Physical Chemistry B,* **113**(32), 11069-11075.

60. Revol, J. F. (1982) On the cross-sectional shape of cellulose crystallites in valonia ventricosa. *Carbohydrate Polymers,* **2**(2), 123-134.

61. Revol, J. F., Godbout, L., Dong, X. M., Gray, D. G., Chanzy, H., and Maret, G. (1994) Chiral nematic suspensions of cellulose crystallites; phase separation and magnetic field orientation. *Liquid Crystals,* **16**(1), 127-134.

62. Kvien, I., and Oksman, K. (2007) Orientation of cellulose nanowhiskers in polyvinyl alcohol. *Applied Physics A: Materials Science and Processing,* **87**(4), 641-643.

63. Huang, Y., and Paul, D. R. (2007) Effect of molecular weight and temperature on physical aging of thin glassy poly(2,6-Dimethyl-1,4-Phenylene oxide) films. *Journal of Polymer Science, Part B: Polymer physics,* **45**, 1390-1398.

64. Pan, J., Hamad, W., and Straus, S. K. (2010) Parameters affecting the chiral nematic phase of nanocrystalline cellulose films. *Macromolecules,* **43**(8), 3851-3858.

65. Espino-Pérez, E., Bras, J., Ducruet, V., Guinault, A., Dufresne, A., and Domenek, S. (2013). Influence of chemical surface modification of cellulose nanowhiskers on thermal, mechanical, and barrier properties of poly(lactide) based bionanocomposites. *European Polymer Journal,* **49**(10), 3144-3154.

66. Aulin, C., Ahok, S., Josefsson, P., Nishino, T., Hirose, Y., Österberg, M., and Wågberg, L. (2009) Nanoscale cellulose films with different crystallinities and mesostructures - their surface properties and interaction with water. *Langmuir,* **25**(13), 7675-7685.

67. Belbekhouche, S., Bras, J., Siqueira, G., Chappey, C., Lebrun, L., Khelifi, B., Marais, S., and Dufresne, A. (2011) Water sorption behavior and gas barrier properties of cellulose whiskers and microfibrils films. *Carbohydrate Polymers,* **83**(4), 1740-1748.

68. Sanchez-Garcia, M. D., and Lagaron, J. M. (2010) On the use of plant cellulose nanowhiskers to enhance the barrier properties of poly-

lactic acid. *Cellulose,* **17**(5), 987-1004.
69. Paralikar, S. A., Simonsen, J., and Lombardi, J. (2008) Poly(vinyl Alcohol)/cellulose nanocrystal barrier membranes. *Journal of Membrane Science,* **320**(1-2), 248-258.
70. Bendahou, A., Kaddami, H., Espuche, E., Gouanvé, F., and Dufresne, A. (2011) Synergism effect of montmorillonite and cellulose whiskers on the mechanical and barrier properties of natural rubber composites. *Macromolecular Materials and Engineering,* **296**(8), 760-769.
71. Ma, H., Burger, C., Hsiao, B. S., and Chu, B. (2011) Ultrafine polysaccharide nanofibrous membranes for water purification. *Biomacromolecules,* **12**(4), 970-976.
72. Qiao, C., Chen, G., Zhang, J., and Yao, J. (2016) Structure and rheological properties of cellulose nanocrystals suspension. *Food Hydrocolloids,* **55**, 19-25.
73. Shafiei-Sabet, S., Hamad, W. Y., and Hatzikiriakos, S. G. (2012) Rheology of nanocrystalline cellulose aqueous suspensions. *Langmuir,* **28**(49), 17124-17133.
74. Dong, X. M., Kimura, T., Revol, J.-F., and Gray, D. G. (1996) Effects of ionic strength on the isotropic-chiral nematic phase transition of suspensions of cellulose crystallites. *Langmuir,* **12**(8), 2076-2082.
75. Ureña-Benavides, E. E., Ao, G., Davis, V. A., and Kitchens, C. L. (2011) Rheology and phase behavior of lyotropic cellulose nanocrystal suspensions. *Macromolecules,* **44**(22), 8990-8998.
76. Li, M. C., Wu, Q., Song, K., Lee, S., Qing, Y., and Wu, Y. (2015) Cellulose nanoparticles: structure-morphology-rheology relationships. *ACS Sustainable Chemistry & Engineering,* **3**(5), 821-832.
77. Ching, Y. C., Ershad Ali, M., Abdullah, L. C., Choo, K. W., Kuan, Y. C., Julaihi, S. J., Chuah, C. H., and Liou, N. S. (2016) Rheological properties of cellulose nanocrystal-embedded polymer composites: A review. *Cellulose,* **23**(2), 1011-1030.
78. Wu, Q., Meng, Y., Wang, S., Li, Y., Fu, S., Ma, L., and Harper, D. (2014) Rheological behavior of cellulose nanocrystal suspension: influence of concentration and aspect ratio. *Journal of Applied Polymer Science,* **131**(15), 1-8.
79. Araki, J., Wada, M., Kuga, S., and Okano, T. (1999) influence of surface charge on viscosity behavior of cellulose microcrystal suspension. *Journal of Wood Science,* **45**(3), 258-261.
80. Boluk, Y., Lahiji, R., Zhao, L., and McDermott, M. T. (2011) Suspension viscosities and shape parameter of cellulose nanocrystals (CNC). *Colloids and Surfaces A: Physicochemical and Engineering Aspects,* **377**(1-3), 297-303.
81. Hasani, M., Cranston, E. D., Westman, G., and Gray, D. G. (2008) Cationic surface functionalization of cellulose nanocrystals. *Soft Matter,* **4**(11), 2238-2244.

82. Ten, E., Bahr, D. F., Li, B., Jiang, L., and Wolcott, M. P. (2012) Effects of cellulose nanowhiskers on mechanical, dielectric, and rheological properties of poly(3-hydroxybutyrate-co-3-hydroxyvalerate)/cellulose nanowhisker composites. *Industrial and Engineering Chemistry Research,* **51**(7), 2941-2951.

83. Durmus, A., Kasgoz, A., and Macosko, C. W. (2007) Linear low density polyethylene (LLDPE)/clay nanocomposites. part i: structural characterization and quantifying clay dispersion by melt rheology. *Polymer,* **48**(15), 4492-4502.

84. Zhou, C., Chu, R., Wu, R., and Wu, Q. (2011) Electrospun polyethylene oxide/cellulose nanocrystal composite nanofibrous mats with homogeneous and heterogeneous microstructures. *Biomacromolecules,* **12**(7), 2617-2625.

85. Littunen, K., Hippi, U., Saarinen, T., and Seppälä, J. (2013) Network formation of nanofibrillated cellulose in solution blended poly(methyl Methacrylate) composites. *Carbohydrate Polymers,* **91**(1), 183-190.

86. Belgacem, M. N., and Gandini, A. (2005) The surface modification of cellulose fibres for use as reinforcing elements in composite materials. *Composite Interfaces,* **12**(1-2), 41-75.

87. John, J. M., and Anandjiwala, R. D. (2008) Recent development in chemical modification and characterization of natural fibre reinforced composites. *Polymer Composites,* **10**, 187-207.

88. Bondeson, D., and Oksman, K. (2007) Dispersion and characteristics of surfactant modified cellulose whiskers nanocomposites. *Composite Interfaces,* **14**(7-9), 617-630.

89. Petersson, L., Kvien, I., and Oksman, K. (2007) Structure and thermal properties of poly(lactic acid)/cellulose whiskers nanocomposite materials. *Composites Science and Technology,* **67**(11-12), 2535-2544.

90. Eyley, S., and Thielemans, W. (2014) Surface modification of cellulose nanocrystals. *Nanoscale,* **6**(14), 7764-7779.

91. Lam, E., Male, K. B., Chong, J. H., Leung, A. C. W., and Luong, J. H. T. (2012) Applications of functionalized and nanoparticle-modified nanocrystalline cellulose. *Trends in Biotechnology,* **30**(5), 283-290.

92. Ávila Ramírez, J. A., Fortunati, E., Kenny, J. M., Torre, L., and Foresti, M. L. (2017) Simple citric acid-catalyzed surface esterification of cellulose nanocrystals. *Carbohydrate Polymers,* **157**, 1358-1364.

93. Zhou, L., He, H., Li, M. C., Huang, S., Mei, C., and Wu, Q. (2018) Grafting polycaprolactone diol onto cellulose nanocrystals via click chemistry: enhancing thermal stability and hydrophobic property. *Carbohydrate Polymers,* **189**, 331-341.

94. Khanjanzadeh, H., Behrooz, R., Bahramifar, N., Gindl-altmutter, W., Bacher, M., Edler, M., and Griesser, T. (2018) Surface chemical functionalization of cellulose nanocrystals. *International Journal of*

Biological Macromolecules, **106**, 1288-1296.

95. Dufresne, A. (2013) Nanocellulose: A new ageless bionanomaterial. *Materials Today*, **16**(6), 220-227.

96. Siqueira, G., Bras, J., and Dufresne, A. (2010) New process of chemical grafting of cellulose nanoparticles with a long chain isocyanate. *Langmuir*, **26**(1), 402-411.

97. Azizi Samir, M. A. S., Alloin, F., Sanchez, J. Y., El Kissi, N., and Dufresne, A. (2004) Preparation of cellulose whiskers reinforced nanocomposites from an organic medium suspension. *Macromolecules*, **37**(4), 1386-1393.

98. van der Berg, O., Capadona, J. R., and Weder, C. (2007) Preparation of homogeneous dispersions of tunicate cellulose whiskers in organic solvents. *Biomacromolecules*, **8**(4), 1353-1357.

99. Habibi, Y., and Dufresne, A. (2008) Highly filled bionanocomposites from functionalized polysaccharides nanocrystals. *Biomacromolecules*, **9**(7), 1974-1980.

100. de Menezes, A. J., Siqueira, G., Curvelo, A. A. S., and Dufresne, A. (2009) Extrusion and characterization of functionalized cellulose whiskers reinforced polyethylene nanocomposites. *Polymer*, **50**(19), 4552-4563.

101. Lin, N., and Dufresne, A. (2013) Physical and/or chemical compatibilization of extruded cellulose nanocrystal reinforced polystyrene nanocomposites. *Macromolecules*, **46**(14), 5570-5583.

102. Hajji, P., Cavaille, J. Y., Favier, V., Gauthier, C., and Vigier, G. (1996) Tensile behavior of nanocomposites from latex and celluose whiskers, *Polymer Composites*, **17**(4), 612-619.

103. Dufresne, A. (2010) Processing of polymer nanocomposites reinforced with polysaccharide nanocrystals. *Molecules*, **15**(6), 4111-4128.

104. Zoppe, J. O., Peresin, M. S., Habibi, Y., Venditti, R. A., and Rojas, O. J. (2009) Reinforcing poly(caprolactone) nanofibers with cellulose nanocrystals. *ACS Applied Materials & Interfaces*, **1**(9), 1996-2004.

105. Magalhães, W. L. E., Cao, X., and Lucia, L. A. (2009) Cellulose nanocrystals/cellulose core-in-shell nanocomposite assemblies. *Langmuir*, **25**(22), 13250-13257.

106. Podsiadlo, P., Choi, S. Y., Shim, B., Lee, J., Cuddihy, M., and Kotov, N. A. (2005) Molecularly engineered nanocomposites: layer-by-layer assembly of cellulose nanocrystals. *Biomacromolecules*, **6**(6), 2914-2918.

107. Jean, B., Heux, L., Dubreuil, F., Chambat, G., and Cousin, F. (2009) Non-electrostatic building of biomimetic cellulose-xyloglucan multilayers. *Langmuir*, **25**(7), 3920-3923.

108. Zhou, C., Wu, Q., Yue, Y., and Zhang, Q. (2011) Application of rod-shaped cellulose nanocrystals in polyacrylamide hydrogels. *Journal of Colloid and Interface Science*, **353**(1), 116-123.

109. Muller, D., Rambo, C. R., Porto, L. M., Schreiner, W. H., and Barra, G. M. O. (2013) Structure and properties of polypyrrole/bacterial cellulose nanocomposites. *Carbohydrate Polymers,* **94**(1), 655-662.
110. Capadona, J. R., Shanmuganathan, K., Tyler, D. J., Rowan, S. J., and Weder, C. (2008) Stimuli-responsive polymer nanocomposites inspired by the sea cucumber dermis. *Science,* **319**, 1370-1375.
111. Brinchi, L., Cotana, F., Fortunati, E., and Kenny, J. M. (2013) Production of nanocrystalline cellulose from lignocellulosic biomass. *Carbohydrate Polymers*, **94**(6), 154-169.
112. Rhim, J. W., Park, H. M., and Ha, C. S. (2013) Bio-nanocomposites for food packaging applications. *Progress in Polymer Science*, **38**(10-11), 1629-1652.
113. Avella, M., De Vlieger, J. J., Errico, M. E., Fischer, S., Vacca, P., and Volpe, M. G. (2005) Biodegradable starch/clay nanocomposite films for food packaging applications. *Food Chemistry*, **93**(3), 467-474.
114. Youssef, A. M., and El-Sayed, S. M. (2018) Bionanocomposites materials for food packaging applications: Concepts and future outlook. *Carbohydrate Polymers,* **193**, 19-27.
115. Elazzouzi-Hafraoui, S., Nishiyama, Y., Putaux, J. L., Heux, L., Dubreuil, F., and Rochas, C. (2008) The shape and size distribution of crystalline nanoparticles prepared by acid hydrolysis of native cellulose. *Biomacromolecules,* **9**(1), 57-65.
116. Khan, A., Khan, R. A., Salmieri, S., Le Tien, C., Riedl, B., Bouchard, J., Chauve, G., Tan, V., and Kamal, M. R., Lacroix, M. (2012) Mechanical and barrier properties of nanocrystalline cellulose reinforced chitosan based nanocomposite films. *Carbohydrate Polymers*, **90**(4), 1601-1608.
117. Huq, T., Salmieri, S., Khan, A., Khan, R. A., Le Tien, C., Riedl, B., Fraschini, C., Bouchard, J., Uribe-Calderon, J., and Kamal, M. R., et al. (2012) Nanocrystalline cellulose (NCC) reinforced alginate based biodegradable nanocomposite film. *Carbohydrate Polymers,* **90**(4), 1757-1763.
118. Li, W., Yue, J., and Liu, S. (2012) Preparation of nanocrystalline cellulose via ultrasound and its reinforcement capability for poly(vinyl alcohol) composites. *Ultrasonics Sonochemistry*, **19**(3), 479-485.
119. Dubief, D., Samain, E., and Dufresne, A. (1999) Polysaccharide microcrystals reinforced amorphous poly(β-Hydroxyoctanoate) nanocomposite materials. *Macromolecules*, **32**(18), 5765-5771.
120. Chazeau, L., Cavaille, J. Y., Canova, G., Dendievel, R., and Boutherin, B. (1999) Viscoelastic properties of plasticized pvc reinforced with cellulose whiskers. *Journal of Applied Polymer Science*, **71**(11), 1797-1808.
121. Yu, H., Yan, C., and Yao, J. (2014) Fully biodegradable food packag-

ing materials based on functionalized cellulose nanocrystals/poly(3-hydroxybutyrate-co-3-hydroxyvalerate) nanocomposites. *RSC Advances*, **4**(104), 59792-59802.

122. Fortunati, E., Armentano, I., Zhou, Q., Iannoni, A., Saino, E., Visai, L., Berglund, L. A., and Kenny, J. M. (2012) Multifunctional bionanocomposite films of poly(lactic acid), cellulose nanocrystals and silver nanoparticles. *Carbohydrate Polymers*, **87**(2), 1596-1605.

123. Batmaz, R., Mohammed, N., Zaman, M., Minhas, G., Berry, R. M., Tam, K. C. (2014) Cellulose nanocrystals as promising adsorbents for the removal of cationic dyes. *Cellulose*, **21**(3), 1655-1665.

124. Sharma, P., Kaur, H., Sharma, M., and Sahore, V. (2011) A review on applicability of naturally available adsorbents for the removal of hazardous dyes from aqueous waste. *Environmental Monitoring and Assessment*, **183**(1-4), 151-195.

125. Crini, G. (2006) Non-conventional low-cost adsorbents for dye removal: A review. *Bioresource Technology*, **97**(9), 1061-1085.

126. Liu, P., Sehaqui, H., Tingaut, P., Wichser, A., Oksman, K., and Mathew, A. P. (2014) Cellulose and chitin nanomaterials for capturing silver ions (Ag^+) from water via surface adsorption. *Cellulose*, **21**(1), 449-461.

127. Abou-Zeid, R. E., Khiari, R., El-Wakil, N., and Dufresne, A. (2018) Current state and new trends in the use of cellulose nanomaterials for wastewater treatment. *Biomacromolecules*, doi: 10.1021/acs.biomac.8b00839.

128. Kardam, A., Raj, K. R., Srivastava, S., and Srivastava, M. M. (2014) Nanocellulose fibers for biosorption of cadmium, nickel, and lead ions from aqueous solution. *Clean Technologies and Environmental Policy*, **16**(2), 385-393.

129. Sirviö, J. A., Hasa, T., Leiviskä, T., Liimatainen, H., and Hormi, O. (2016) Bisphosphonate nanocellulose in the removal of vanadium(V) from water. *Cellulose*, **23**(1), 689-697.

130. Araki, J., Wada, M., and Kuga, S. (2001) Steric stabilization of a cellulose microcrystal suspension by poly(ethylene glycol) grafting. *Langmuir*, **17**(1), 21-27.

131. Liu, P., Borrell, P. F., Božič, M., Kokol, V., Oksman, K., and Mathew, A. P. (2015) nanocelluloses and their phosphorylated derivatives for selective adsorption of Ag^+, Cu^{2+} and Fe^{3+} from industrial effluents. *Journal of Hazardous Materials*, **294**, 177-185.

132. Qiao, H., Zhou, Y., Yu, F., Wang, E., Min, Y., Huang, Q., Pang, L., and Ma, T. (2015) Effective removal of cationic dyes using carboxylate-functionalized cellulose nanocrystals. *Chemosphere*, **141**, 297-303.

133. Yu, X., Tong, S., Ge, M., Wu, L., Zuo, J., Cao, C., and Song, W. (2013) Adsorption of heavy metal ions from aqueous solution by carboxylated cellulose nanocrystals. *Journal of Environmental Sciences (China)*, **25**(5), 933-943.

134. Singh, K., Arora, J. K., Sinha, T. J. M., and Srivastava, S. (2014) Functionalization of nanocrystalline cellulose for decontamination of cr(iii) and cr(vi) from aqueous system: computational modeling approach. *Clean Technologies and Environmental Policy*, **16**(6), 1179-1191.

135. Braun, B., Dorgan, J. R., Hollingsworth, L. O. (2012) Supramolecular ecobionanocomposites based on polylactide and cellulosic nanowhiskers: synthesis and properties. *Biomacromolecules*, **13**(7), 2013-2019.

136. He, X., Male, K. B., Nesterenko, P. N., Brabazon, D., Paull, B., and Luong, J. H. T. (2013) Adsorption and desorption of methylene blue on porous carbon monoliths and nanocrystalline cellulose. *ACS Applied Materials & Interfaces*, **5**(17), 8796-8804.

137. Eyley, S., and Thielemans, W. (2011) Imidazolium grafted cellulose nanocrystals for ion exchange applications. *Chemical Communications*, **47**(14), 4177-4179.

138. Mohammed, N., Grishkewich, N., Tam, K. C., and Berry, R. (2016) Pristine and Surface Functionalised Cellulose Naocrystals (Cncs) Incorporated Hydrogel Beads And Uses Thereof, patent US2016/0175812A1.

139. Jin, L., Sun, Q., Xu, Q., and Xu, Y. (2015) Adsorptive removal of anionic dyes from aqueous solutions using microgel based on nanocellulose and polyvinylamine. *Bioresource Technology*, **197**, 348-355.

140. Jin, L., Li, W., Xu, Q., and Sun, Q. (2015) Amino-functionalized nanocrystalline cellulose as an adsorbent for anionic dyes. *Cellulose*, **22**(4), 2443-2456.

141. Mohammed, N., Grishkewich, N., Berry, R. M., and Tam, K. C. (2015) Cellulose nanocrystal-alginate hydrogel beads as novel adsorbents for organic dyes in aqueous solutions. *Cellulose*, **2**(6), 3725-3738.

142. Wang, X., Yao, C., Wang, F., and Li, Z. (2017) Cellulose-based nanomaterials for energy applications. *Small*, **13**(42), 1-19.

143. Hoeng, F., Denneulin, A., and Bras, J. (2016,) Use of nanocellulose in printed electronics: A review. *Nanoscale* **8**(27), 13131-13154.

144. van den Berg, O., Schroeter, M., Capadona, J. R., and Weder, C. (2007) Nanocomposites based on cellulose whiskers and semiconducting conjugated polymers. *Journal of Materials Chemistry*, **17**(26), 2746-2753.

145. Hamad, W. Y., and Atifi, S. (2016) Flexible, Semiconducting Nanocomposite Materials Based On Nanocrystalline Cellulose And Polyalanine, patent US9384867B2.

146. Xiong, R., Hu, K., Grant, A. M., Ma, R., Xu, W., Lu, C., Zhang, X., and Tsukruk, V. V. (2016) Ultrarobust transparent cellulose nanocrystal-graphene membranes with high electrical conductivity. *Advanced Materials*, **28**(7), 1501-1509.

147. Sasso, C., Zeno, E., Petit-Conil, M., Chaussy, D., Belgacem, M. N., Tapin-Lingua, S., and Beneventi, D. (2010) Highly conducting polypyrrole/cellulose nanocomposite films with enhanced mechanical properties. *Macromolecular Materials and Engineering*, **295**(10), 934-941.

148. Hu, L., Zheng, G., Yao, J., Liu, N., Weil, B., Eskilsson, M., Karabulut, E., Ruan, Z., Fan, S., and Bloking, J. T., et al. (2013) Transparent and conductive paper from nanocellulose fibers. *Energy and Environmental Science*, **6**(2), 513-518.

149. Valentini, L., Cardinali, M., Fortunati, E., Torre, L., and Kenny, J. M. (2013) A novel method to prepare conductive nanocrystalline cellulose/graphene oxide composite films. *Materials Letters*, **105**, 4-7.

150. Dang, L. N., and Seppälä, J. (2015) Electrically conductive nanocellulose/graphene composites exhibiting improved mechanical properties in high-moisture condition. *Cellulose*, **22**(3), 1799-1812.

151. Tang, Y., He, Z., Mosseler, J. A., and Ni, Y. (2014) Production of highly electro-conductive cellulosic paper via surface coating of carbon nanotube/graphene oxide nanocomposites using nanocrystalline cellulose as a binder. *Cellulose*, **21**(6), 4569-4581.

152. Szcześniak, L., Rachocki, A., and Tritt-Goc, J. (2008) Glass transition temperature and thermal decomposition of cellulose powder. *Cellulose*, **15**(3), 445-451.

153. Angelo, P. D., Cole, G. B., Sodhi, R. N., and Farnood, R. R. (2012) Conductivity of inkjet-printed PEDOT:PSS-SWCNTs on uncoated papers. *Nordic Pulp and Paper Research Journal*, **27**(02), 486-495.

154. Wood, H. L. K., Pekarovic, J., Pekarovicova, A., Fleming, P. D., and Bliznyuk, V, E. (2005) The properties of conducting polymers and substrates for printed electronics. *IS&T Digital Fabrication*, 197-202.

155. Bollström, R., Pettersson, F., Dolietis, P., Preston, J., Österbacka, R., and Toivakka, M. (2014) Impact of humidity on functionality of on-paper printed electronics. *Nanotechnology*, **25**(9), 1-12.

156. Hoeng, F., Bras, J., Gicquel, E., Krosnicki, G., and Denneulin, A. (2017) Inkjet printing of nanocellulose-silver ink onto nanocellulose coated cardboard. *RSC Advances*, **7**(25), 15372-15381.

157. Liew, S. Y., and Walsh, D. A., Thielemans, W. (2013) High total-electrode and mass-specific capacitance cellulose nanocrystal-polypyrrole nanocomposites for supercapacitors. *RSC Advances*, **3**(24), 9158-9162.

158. Yang, X., Shi, K., Zhitomirsky, I., and Cranston, E. D. (2015) Cellulose nanocrystal aerogels as universal 3D lightweight substrates for supercapacitor materials. *Advanced Materials*, **27**(40), 6104-6109.

159. Chen, W., Yu, H., Lee, S. Y., Wei, T., Li, J., Fan, Z. (2018) Nanocellu-

lose: a promising nanomaterial for advanced electrochemical energy storage. *Chemical Society Reviews,* **47**(8), 2837-2872.
160. Cho, T. H., Tanaka, M., Onishi, H., Kondo, Y., Nakamura, T., Yamazaki, H., Tanase, S., and Sakai, T. (2008) Battery performances and thermal stability of polyacrylonitrile nano-fiber-based nonwoven separators for Li-ion battery. *Journal of Power Sources,* **181**(1), 155-160.
161. de Rodriguez, N. L. G., Thielemans, W., and Dufresne, A. (2006) Sisal cellulose whiskers reinforced polyvinyl acetate nanocomposites. *Cellulose,* **13**(3), 261-270.
162. Samir, M. A. S. A., Alloin, F., Sanchez, J. Y., and Dufresne, A. (2004) Cross-linked nanocomposite polymer electrolytes reinforced with cellulose whiskers. *Macromolecules,* **37**(13), 4839-4844.
163. Lalia, B. S., Samad, Y. A., and Hashaikeh, R. (2013) Nanocrystalline cellulose-reinforced composite mats for lithium-ion batteries: electrochemical and thermomechanical performance. *Journal of Solid State Electrochemistry,* **17**(3), 575-581.
164. Bolloli, M., Antonelli, C., Molméret, Y., Alloin, F., Iojoiu, C., and Sanchez, J. Y. (2016) Nanocomposite poly(vynilidene fluoride)/nanocrystalline cellulose porous membranes as separators for lithium-ion batteries. *Electrochimica Acta,* **214**, 38-48.
165. Halib, N., Perrone, F., Cemazar, M., Dapas, B., Farra, R., Abrami, M., Chiarappa, G., Forte, G., Zanconati, F., and Pozzato, G., Murena, N., Lapasin, R., Cansolino, L., Grassi, G., and Grassi, M. (2017) Potential applications of nanocellulose-containing materials in the biomedical field. *Materials,* **10**(8), 1-31.
166. Li, W., Guo, R., Lan, Y., Zhang, Y., Xue, W., and Zhang, Y. (2014) Preparation and properties of cellulose nanocrystals reinforced collagen composite films. *Journal of Biomedical Materials Research, Part A,* **102**(4), 1131-1139.
167. Lee, J., and Deng, Y. (2012) Increased mechanical properties of aligned and isotropic electrospun PVA nanofiber webs by cellulose nanowhisker reinforcement. *Macromolecular Research,* **20**(1), 76-83.
168. Rueda, L., Saralegi, A., Fernández-d'Arlas, B., Zhou, Q., Alonso-Varona, A., Berglund, L. A., Mondragon, I., Corcuera, M. A., and Eceiza, A. (2013) In situ polymerization and characterization of elastomeric polyurethane-cellulose nanocrystal nanocomposites. cell response evaluation. *Cellulose,* **20**(4), 1819-1828.
169. He, X., Xiao, Q., Lu, C., Wang, Y., Zhang, X., Zhao, J., Zhang, W., Zhang, X., and Deng, Y. (2014) Uniaxially aligned electrospun all-cellulose nanocomposite nanofibers reinforced with cellulose nanocrystals: scaffold for tissue engineering. *Biomacromolecules,* **15**(2), 618-627.
170. Kumar, A., Negi, Y. S., Choudhary, V., and Bhardwaj, N. K. (2014)

Microstructural and mechanical properties of porous biocomposite scaffolds based on polyvinyl alcohol, nano-hydroxyapatite and cellulose nanocrystals. *Cellulose*, **21**(5), 3409-3426.

171. Zhou, C., Shi, Q., Guo, W., Terrell, L., Qureshi, A. T., Hayes, D. J., and Wu, Q. (2013) Electrospun bio-nanocomposite scaffolds for bone tissue engineering by cellulose nanocrystals reinforcing maleic anhydride grafted PLA. *ACS Applied Materials & Interfaces*, **5**(9), 3847-3854.

172. Sinha, A., Martin, E. M., Lim, K.-T., Carrier, D. J., Han, H., Zharov, V. P., and Kim, J.-W. (2015) Cellulose nanocrystals as advanced "green" materials for biological and biomedical engineering. *Journal of Biosystems Engineering*, **40**(4), 373-393.

173. Domingues, R. M. A., Silva, M., Gershovich, P., Betta, S., Babo, P., Caridade, S. G., Mano, J. F., Motta, A., Reis, R. L., and Gomes, M. E. (2015) Development of injectable hyaluronic acid/cellulose nanocrystals bionanocomposite hydrogels for tissue engineering applications. *Bioconjugate Chemistry*, **26**(8), 1571-1581.

174. Yang, J., Han, C. R., Duan, J. F., Ma, M. G., Zhang, X. M., Xu, F., and Sun, R. C. (2013) Synthesis and characterization of mechanically flexible and tough cellulose nanocrystals-polyacrylamide nanocomposite hydrogels. *Cellulose*, **20**(1), 227-237.

175. Yang, X., Bakaic, E., Hoare, T., and Cranston, E. D. (2013) Injectable polysaccharide hydrogels reinforced with cellulose nanocrystals: morphology, rheology, degradation, and cytotoxicity. *Biomacromolecules*, **14**(12), 4447-4455.

176. Wang, H., Qian, C., and Roman, M. (2011) Effects of pH and salt concentration on the formation and properties of chitosan-cellulose nanocrystal polyelectrolyte-macroion complexes. *Biomacromolecules*, **12**(10), 3708-3714.

177. Jackson, J. K., Letchford, K., Wasserman, B. Z., Ye, L., Hamad, W. Y., and Burt, H. M. (2011) The use of nanocrystalline cellulose for the binding and controlled release of drugs. *International Journal of Nanomedicine*, **6**, 321-330.

Index

www.ingramcontent.com/pod-product-compliance
Lightning Source LLC
Chambersburg PA
CBHW050459190326
41458CB00005B/1359